Theo Fett

Stress Intensity Factors – T-Stresses – Weight Functions

Supplement Volume

Schriftenreihe des Instituts für Keramik im Maschinenbau
IKM 55

Institut für Keramik im Maschinenbau,
Karlsruher Institut für Technologie

Stress Intensity Factors
T-Stresses
Weight Functions

Supplement Volume

by
Theo Fett

Institute of Ceramics in Mechanical Engineering (IKM),
Karlsruhe Institute of Technology

Impressum

Karlsruher Institut für Technologie (KIT)
KIT Scientific Publishing
Straße am Forum 2
D-76131 Karlsruhe
www.uvka.de

KIT – Universität des Landes Baden-Württemberg und nationales
Forschungszentrum in der Helmholtz-Gemeinschaft

KIT Scientific Publishing 2009
Print on Demand

ISSN: 1436-3488
ISBN: 978-3-86644-446-1

Preface

In "Stress intensity factors, T-stresses, Weight functions", Volume 50 of this series, predominantly one-dimensional cracks (cracks of constant width) were considered in homogeneous materials.

This supplement volume compiles new results on one-dimensional cracks and results obtained for more complicated crack problems as e.g.

- Straight cracks in dissimilar materials,

- Two-dimensional cracks.

In addition, the inverse weight function problem is briefly addressed that deals with the evaluation of the stresses for the case of known stress intensity factors. Finally, some corrections are given referring to Vol. 50.

The author has to thank his colleagues Gabriele Rizzi (Forschungszentrum Karlsruhe, IMF) for additional Finite Element computations, Michael Politzky (Forschungszentrum Karlsruhe, IKET) and Rainer Müller (IKM, University of Karlsruhe) for support in the field of computer application, and Stefan Fünfschilling (IKM, University of Karlsruhe) for providing experimental R-curve results used in the determination of bridging stresses.

Universität Karlsruhe

Karlsruhe, November 2009   Theo Fett

CONTENTS

Part E

SUPPLEMENTS TO ONE-DIMENSIONAL CRACKS

III

Part F

TWO-DIMENSIONAL CRACKS 79

Part G

CORRECTIONS IN VOLUME 50 133

PART E

SUPPLEMENTS TO

ONE-DIMENSIONAL CRACKS

<u>Part E</u> deals with:

Asymptotic term for the T-stress Green's function

Cracks in bi-materials

Weight function and compliance for cracks ahead of slender notches

A trapezoidal test specimen

Additional results for the DCDC specimen

Slant and kink cracks in finite bars

Inverse weight function problem

E1

Asymptotic term of the Green's function for T

E1.1   Green's function for symmetric crack problems

As had been outlined in Section A1 of [E1.1], the T-stress can be expressed by an integral of the form

$$T = -\sigma_y\big|_{x=a} + \int\limits_0^a t(x,a)\,\sigma(x)\,dx \qquad (E1.1.1)$$

where the integration has to be performed over the stress field σ_y in the uncracked body.

If in the uncracked body a σ_x stress component already exists at the location of the tip of the prospective crack, the total T-value is obtained by adding this stress contribution, i.e.

$$T = \sigma_x\big|_{x=a} - \sigma_y\big|_{x=a} + \int\limits_0^a t(x,a)\,\sigma(x)\,dx \qquad (E1.1.2)$$

In order to describe the Green's function, it is distinguished here between a term t_0 representing the asymptotic limit case of near-tip behaviour and an additive regular term t_{reg} which includes information about the special shape of the component and the finite dimensions,

$$t = t_0 + t_{reg} \qquad (E.1.1.3)$$

E1.2   The asymptotic term

In order to obtain information on the asymptotic behaviour of the weight or Green's function, the near-tip behaviour shall be considered exclusively. Therefore, a small section of the body (dashed circle) very close to the crack tip is taken into consideration (Fig. E1.1). This near-tip zone may be zoomed very strongly. Consequently, the outer borders of the component move to infinity. This results in the case of a semi-infinite crack in an infinite body. If the crack faces are loaded by a couple of forces P at location $x=x_0$ with $a-x_0 \ll a$, the stress state can be described in terms of the Westergaard stress function [E1.2]:

$$Z = \frac{P}{\pi} \frac{1}{z+b} \sqrt{\frac{b}{z}}, \quad z = \xi + iy \tag{E1.2.1}$$

The regular part of to the stress function is $(z, b \neq 0)$

$$Z_{reg} = -\frac{P}{\pi} \frac{1}{z+b} \sqrt{\frac{z}{b}} \tag{E1.2.2}$$

from which the regular part of the x-stress component results as

$$\sigma_x = \operatorname{Re} Z - y \operatorname{Im}(dZ/dz) \quad \Rightarrow \quad \sigma_x\big|_{y=0} = \operatorname{Re}\{Z\}\big|_{y=0} \tag{E1.2.3}$$

$$\sigma_{x,reg}\big|_{y=0} = \operatorname{Re}\{Z_{reg}\}\big|_{y=0} = -\frac{P}{\pi} \frac{\sqrt{x'-a}}{(x'-x_0)\sqrt{a-x_0}}, \quad x' > a \tag{E1.2.4}$$

In eq.(E1.2.4) x' is the x-value at which the constant σ_x-stress term is evaluated. The regular x-stress at $x' = a$ is then given by

$$\sigma_{x,reg}\big|_{x' \to a} = -\frac{P}{\pi} \lim_{x' \to a} \frac{\sqrt{x'-a}}{(x'-x_0)\sqrt{a-x_0}} \tag{E1.2.5}$$

and the Green's function reads

$$\Rightarrow \qquad t_0 = -\frac{1}{\pi} \lim_{x' \to a} \frac{\sqrt{x'-a}}{(x'-x_0)\sqrt{a-x_0}} \ . \tag{E1.2.6}$$

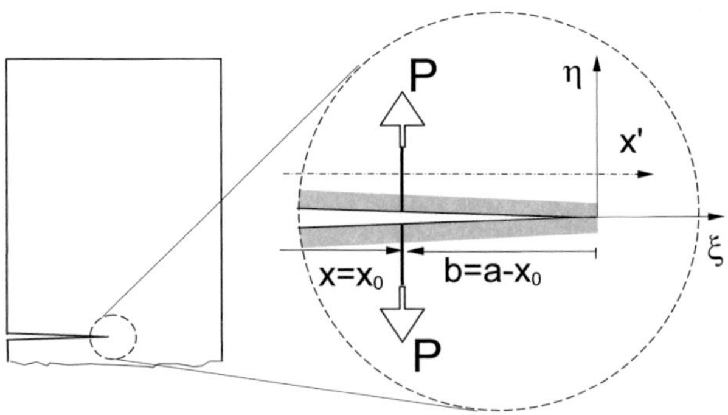

Fig. E1.1 Situation at the crack tip for an asymptotic stress consideration.

From (E1.2.6), the T-stress can be derived for a couple of forces acting on a semi-infinite crack in an infinite body, namely,

$$T = t_0 = \begin{cases} 0 & \text{for } x_0 < a \\ -\infty & \text{for } x_0 = a \end{cases}. \tag{E1.2.7}$$

An arbitrary crack loading σ_y may be represented by a Taylor series expansion at the location of the crack tip

$$\sigma_y(x) = \sigma_y\big|_{x=a} - \frac{d\sigma_y}{dx}\bigg|_{x=a}(a-x) + \frac{1}{2}\frac{d^2\sigma_y}{dx^2}\bigg|_{x=a}(a-x)^2 - + ... \tag{E1.2.8}$$

After replacing x_0 by x, the corresponding T-stress contribution resulting from the asymptotic part of the Green's function reads

$$T = \int_0^a t_0(x',a,x)\,\sigma(x)\,dx = -\frac{1}{\pi}\sigma_y\big|_{x=a}\lim_{x'\to a}\sqrt{x'-a}\int_0^a\frac{dx}{(x'-x)\sqrt{a-x}} \tag{E1.2.9}$$

or in integrated form

$$T = -\frac{1}{\pi}\sigma_y\big|_{x=a}\lim_{x'\to a}\sqrt{x'-a}\left[\frac{2}{\sqrt{x'-a}}\arctan\sqrt{\frac{x'-a}{a-x}}\right]_0^a = -\sigma_y\big|_{x=a} \tag{E1.2.10}$$

The two relations

$$T = t_0 = \begin{cases} 0 & \text{for } x < a \\ -\infty & \text{for } x = a \end{cases}, \qquad T = \int_0^a t_0\,\sigma(x)\,dx = -\sigma_y\big|_{x=a} \tag{E1.2.11}$$

are in agreement with the definitions of the Dirac δ-function. As done in [E1.3] we can finally write for the singular term

$$t_0 = -\delta(a-x) \tag{E1.2.12}$$

It should be mentioned that recently the δ-function behaviour was confirmed also by Chen and Lin [E1.4]. Insertion of (E1.2.12) into the weight function relation yields

$$T = -\sigma_y\big|_{x=a} + \int_0^a t_{reg}\,\sigma(x)\,dx \tag{E1.2.13}$$

These two parts of the T-stress are caused by the crack. In absence of a crack there is of course no x-stress component. If in the uncracked body an σ_x stress component already exists at the location of the tip of the prospective crack, the total T-value is obtained by adding this stress contribution, to eq.(E1.2.13), i.e.

$$T = \sigma_x\big|_{x=a} - \sigma_y\big|_{x=a} + \int_0^a t_{reg}\,\sigma(x)\,dx \qquad \text{(E1.2.14)}$$

representing the final result.

References E1

E1.1 Fett, T., Stress intensity factors, T-stresses, Weight functions, IKM 50, Universitätsverlag Karls-ruhe, 2008.

E1.2 Sham, T.L., The theory of higher order weight functions for linear elastic plane problems, Int. J. Solids and Struct. **25**(1989), 357-380.

E1.3 Fett, T., Rizzi, G., T-stress solutions determined by Finite Element computations, Report FZKA 6937, Forschungszentrum Karlsruhe, Karlsruhe 2004.

E1.4 Chen, Y.Z., Lin, X.Y., Comments on "Approximate Green's functions for singular and higher order terms of an edge crack in a finite plate" by Xiao and Karihaloo [Engng Fract Mech 2002;69:959–981], Engng. Fract. Mech. **75**(2008), 4844-48.

E2

Stress terms in dissimilar materials

E2.1 Basic relations

E2.1.1 Interface crack

The mechanical behaviour of a bi-material joint (consisting of materials "1" and "2") is characterised by the Dundurs parameters α and β which are defined as

$$\alpha = \frac{\overline{E}_1 - \overline{E}_2}{\overline{E}_1 + \overline{E}_2} \quad , \quad \overline{E} = \frac{E}{1 - v^2} \tag{E2.1.1}$$

$$\beta = \frac{1}{2}\frac{\mu_1(1-2v_2) - \mu_2(1-2v_1)}{\mu_1(1-v_2) + \mu_2(1-v_1)}, \quad \mu = \frac{E}{2(1+v)} \tag{E2.1.2}$$

with Young's modulus E and Poisson's ratio v.

For a crack lying directly on the interface (Fig. E2.1), the stress field is given by a complex interface stress intensity factor K [E2.1] expressed as

$$K = K_1 + iK_2 \tag{E2.1.3}$$

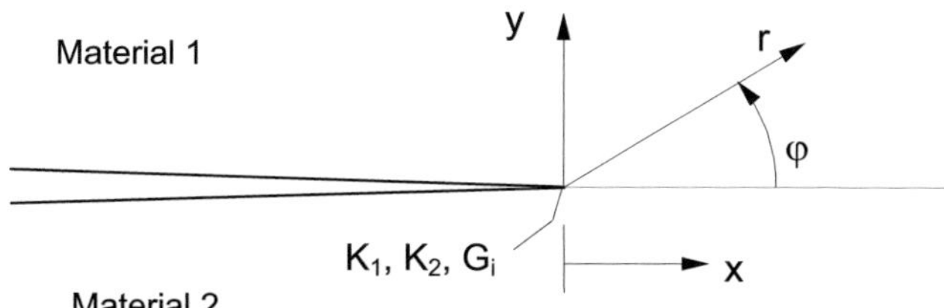

Fig. E2.1 Interface crack (geometric data).

The full stress solution for such cracks was given by Sih and Chen [E2.2]. The tractions on the interface ahead of the crack tip ($r{\to}x$, $\varphi{=}0$) are

$$\sigma_{22} + i\sigma_{12} = \frac{K}{\sqrt{2\pi x}}x^{i\varepsilon} \tag{E2.1.4}$$

with

$$\varepsilon = \frac{1}{2\pi} \ln \frac{1-\beta}{1+\beta} \qquad (E2.1.5)$$

The energy release rate G of the crack advancing in the interface direction reads

$$G = \frac{K_1^2 + K_2^2}{E^*} \qquad (E2.1.6)$$

with the effective modulus E^* defined as

$$\frac{1}{E^*} = \frac{1}{2}\left(\frac{1}{\overline{E}_1} + \frac{1}{\overline{E}_2}\right)\frac{1}{\cosh^2 \pi\varepsilon} \qquad (E2.1.7)$$

As can be seen from (E2.1.6) and (E2.1.7), the energy release rate can be applied to interface cracks in the same way as to homogeneous materials. If the stresses in the vicinity of the crack tip are of interest, the computations are much more complicated. Knowledge of these stresses is necessary to decide whether a crack will extend in its initial direction (i.e. on the interface) or kink into one of the two materials [E2.3]. When β=0 and, consequently, ε=0, the stress intensity factors K_1 and K_2 can be interpreted as conventional stress intensity factors K_I and K_{II} ($K_1 \rightarrow K_I$, $K_2 \rightarrow K_{II}$).

E2.1.2 Kink crack

The conditions of kinking are outlined in detail in the papers of He and Hutchinson [E2.3] and He et al.[E2.4]. The stress intensity factors of the kinked crack (Fig. E2.2) k_I and k_{II} are conventional stress intensity factors, because the crack tip now is surrounded by one material exclusively. These stress intensity factors are related to the stress intensity factors of the unkinked crack as well as to the constant stress term σ_0 in the material in which the crack will kink

$$k_I + ik_{II} = cK\ell^{i\varepsilon} + \overline{dK}\ell^{-i\varepsilon} + b\sigma_0\sqrt{\ell} \qquad (E2.1.8)$$

where c, d, and b are dimensionless complex parameters depending on the Dundurs parameters and the kink angle ω. In the case of Fig. E2.2, the relevant constant stress term is σ_{02}. For homogeneous materials the stress σ_0 is identical with the so-called T-stress. The effects of T on path stability under mixed-mode loading were discussed in detail by Cotterell and Rice [E2.5].
For the special case of β=0, the stress intensity factors of the kink are

$$k_I = (c_R + d_R)K_1 - (c_I + d_I)K_2 + b_1\sigma_0\sqrt{\ell} \qquad (E2.1.9)$$

$$k_{II} = (c_I - d_I)K_1 + (c_R - d_R)K_2 + b_2\sigma_0\sqrt{\ell} \qquad (E2.1.10)$$

The energy release rate G_k of the kink crack then results from

$$G_k = \frac{k_I^2 + k_{II}^2}{\overline{E}_2} \qquad (E2.1.11)$$

The a priori unknown kink angle ω can be determined from eq.(E2.1.11) taking $\ell \to 0$. The ratio between the interface energy release rate G_i and the maximum value of $G_k(\omega)$ under kink conditions, $G_{k,max}$, have to be determined. The value of $G_i/G_{k,max}$ can then be compared with the ratio of the mode-dependent interface toughness Γ_i and the toughness of the material in which the crack kinks, Γ_k, i.e. Γ_i/Γ_k. This allows deciding whether the crack is able to kink (for details see e.g. [E2.4]).

In order to model the crack growth and kink behaviour of interface cracks, it is necessary to determine the stress intensity factors K_1, K_2 (or K_I and K_{II}, for $\beta = 0$) and the constant stress terms for the specimens of interest. Whereas the stress intensity factors are available for a large number of infinite and semi-infinite bodies (see e.g.[E2.6]), there is experimental interest in practically used test specimens. This especially holds for the constant x-stress terms.

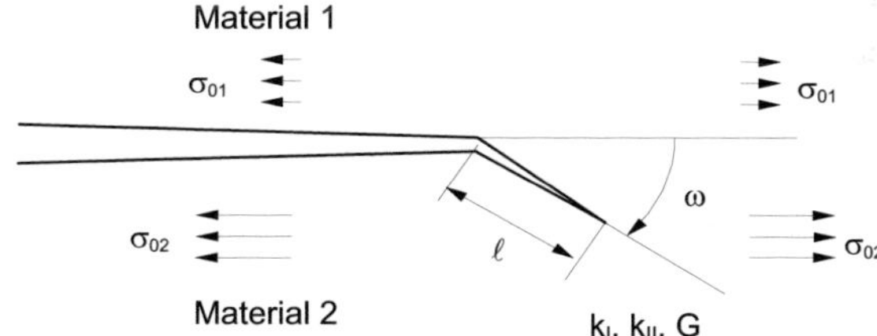

Fig. E2.2 Geometry of a kinked crack with constant stress terms of the initial (unkinked) crack.

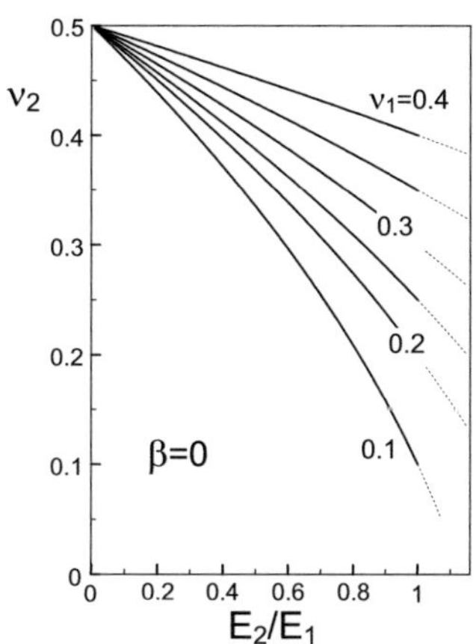

Fig. E2.3 Poisson's ratio ν_2 required for a disappearing Dundurs parameter β.

As emphasized by Hutchinson [E2.7], *"the clarity in interpretation achieved by taking β to be zero is often worth the small sacrifice in accuracy"*.

Having this in mind, the special case of β=0 will be considered in detail below. At a given ratio of Young's moduli E_2/E_1 and a prescribed Poisson ratio ν_1, the second Poisson ratio that fulfils β=0 is given as

$$\nu_2 = -\tfrac{1}{4} + \sqrt{\tfrac{9}{16} - \tfrac{1}{2}(1 - \nu_1 - 2\nu_1^2)E_2/E_1} \qquad (E2.1.12)$$

Figure E2.3 represents this dependency for several values of ν_1.

E2.2 Double Cantilever Beam

The double-cantilever-beam (DCB) specimen is shown in Fig. E2.4. A line load P/B ($B=$ specimen thickness, often chosen as $B=1$) is applied at the end of the cantilever normally to the crack face.

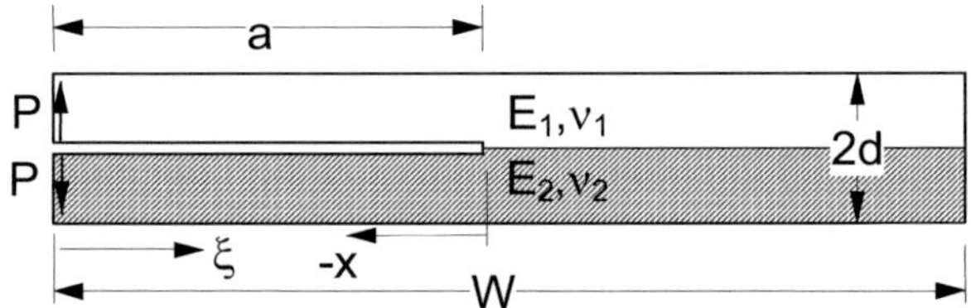

Fig. E2.4 Double-cantilever-beam specimen made of dissimilar materials.

Finite element (FE) computations were carried out with ABAQUS, version 6.2, which provided the stress intensity factors K_I as well as the energy release rate in the form of the J-integral. For the FE computations, the geometry was chosen to be W=6000 and d=500-1500. In total, about 7400 elements with 23000 nodes were used. The crack tip region was modelled with 8-node iso-parametric elements collapsed on one side.

Results for a slender DCB specimen (d/W=12)
The energy release rate was determined as a function of the first Dundurs parameter α. By use of eqs.(E2.1.6) and (E2.1.7), an effective stress intensity factor K_{eff} can be defined as

$$K_{eff} = \sqrt{GE^*} \qquad (E2.2.1)$$

with the effective Young's modulus E^* given by eq.(E2.1.7). The results for a/d=6 are represented in this form in Fig. E2.5. There is no significant dependency on α (this had to be expected). The solid and dashed horizontal lines indicate the average value and the span of data, i.e. $K_{eff}\sqrt{d}/(P/B)$=23.08 (±0.3%). Such small deviations are within the range of accuracy of the FE method.

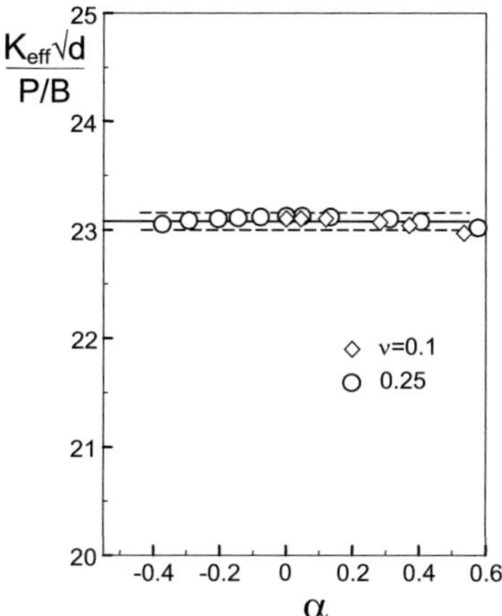

Fig. E2.5 Energy release rate G of an advancing interface crack expressed by the effective stress intensity factor according to eq.(E2.2.1) at a/d=6 and β=0.

Mixed-mode stress intensity factors K_I and K_{II} are plotted in Fig. E2.6 as functions of the modulus ratio E_2/E_1. Maximum K_I and trivially disappearing K_{II} are found for E_2/E_1=1. Only a slight influence of ν_1 is visible. Figure E2.7 shows similar plots for the dependency on the Dundurs parameter α.

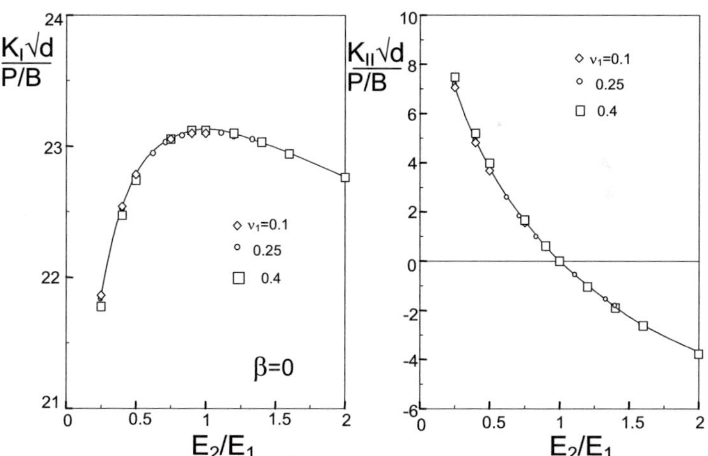

Fig. E2.6 Stress intensity factor contributions for a/d=6 (β=0).

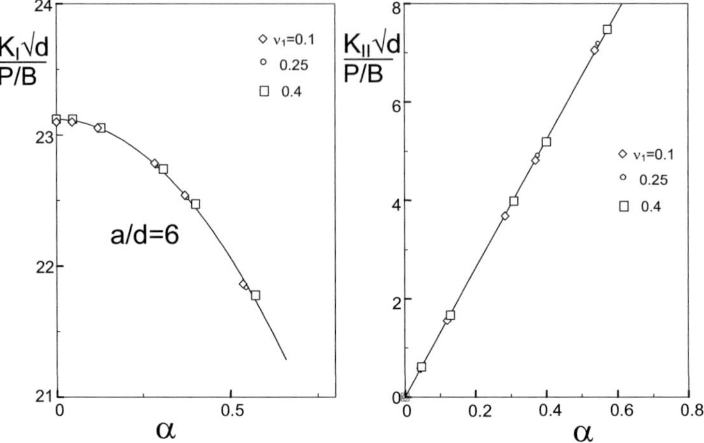

Fig. E2.7 Stress intensity factors K_I and K_{II} as functions of the first Dundurs parameter α for a/d=6 (β=0).

In homogeneous materials, only one constant stress term exists, the T-stress. This value can be determined easily from the x-stresses at the free crack surfaces, as there is no other stress component near the crack tip. Moreover, the ABAQUS, version 6.2 directly provides T.

In the case of an interface crack, two different values exist for the constant x-stress term, here denoted as σ_{01} for material "1" and σ_{02} for material "2". Their determination requires a least-squares evaluation procedure. Whereas for pure mode-I stress fields the singular stresses vanish at $\varphi=\pm\,\pi$, the mode-II stress intensity factor yields singular x-stresses also at the crack surface under mixed-mode conditions. The total stresses caused by the mode-II stress intensity factor and the constant stress terms are

$$\sigma_x = -\frac{K_{II}}{\sqrt{2r\pi}}(2+\cos\tfrac{1}{2}\varphi\cos\tfrac{3}{2}\varphi)\sin\tfrac{1}{2}\varphi+\sigma_0 \tag{E2.2.2}$$

The x-stresses at the crack faces ($\varphi=\pm\pi$) read

$$\sigma_x = -\frac{2K_{II}}{\sqrt{-2x\pi}}+\sigma_0 \tag{E2.2.3}$$

For the evaluation of σ_0, eq.(E2.2.3) may be rewritten as

$$\sigma_x\sqrt{-x} = -\frac{\sqrt{2}K_{II}}{\sqrt{\pi}}+\sigma_0\sqrt{-x} \tag{E2.2.4}$$

The constant stress is then determined from the slope of a $\sigma_x\sqrt{(-x)}$ versus $\sqrt{(-x)}$ plot, as shown in Fig. E2.8 for $a/d=6$ and two different ratios of Young's moduli. The values at $\sqrt{(-x)}=0$ provide the mode-II stress intensity factor.

Figure E2.9 shows the constant stress terms as functions of the Young's modulus ratio E_2/E_1. A plot of σ_0 versus the Dundurs parameter α is shown in Fig. E2.10. A straight-line behaviour can be concluded.

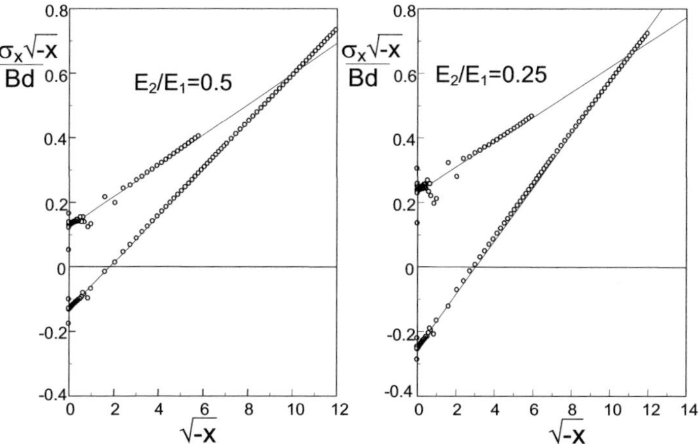

Fig. E2.8 Determination of the constant stress terms from the slope of the straight lines ($\beta=0$).

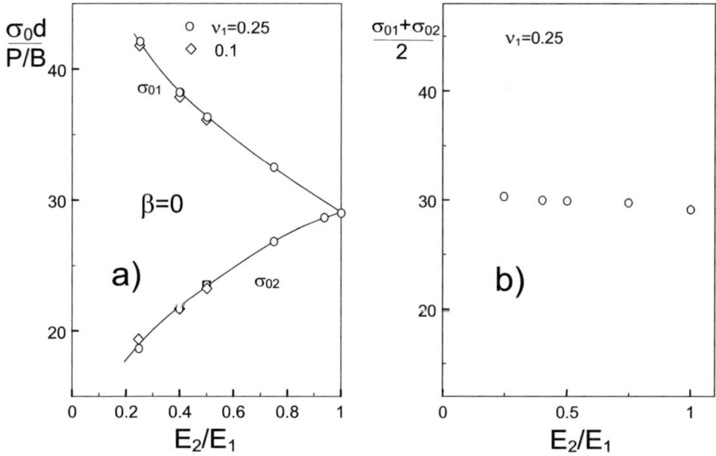

Fig. E2.9 a) Constant stress terms versus ratio E_2/E_1, b) average of the two constant stress values σ_{01} and σ_{02} ($a/d=6$, $\beta=0$).

13

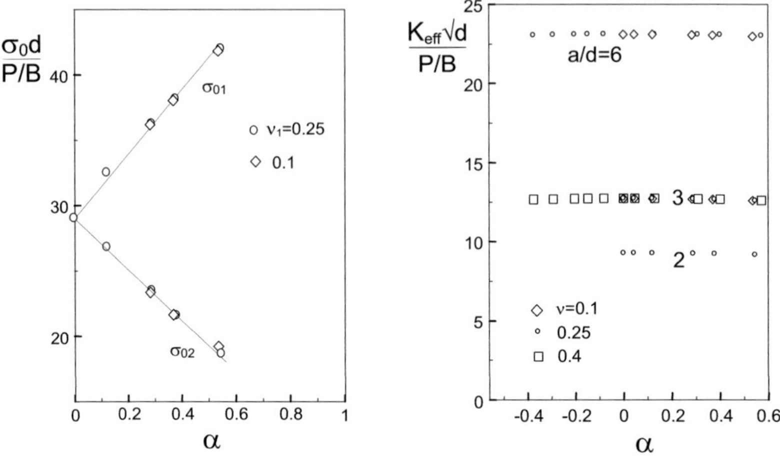

Fig. E2.10 a) Constant stress terms σ_{01} and σ_{02} as functions of the first Dundurs parameter α for a/d=6, β=0, b) Influence of the a/d ratio on the effective stress intensity factor.

Figure E2.10b shows the effective stress intensity factor according to eq.(E2.1.13) for three different a/d ratios. The effective stress intensity factor increases with a/d, but is nearly independent of the parameters α and ν_1.

The stress intensity factor contributions K_I and K_{II} are plotted in Fig. E2.11 as functions of the Dundurs parameter α, the ratio a/d, and the Poisson ratio ν_1.

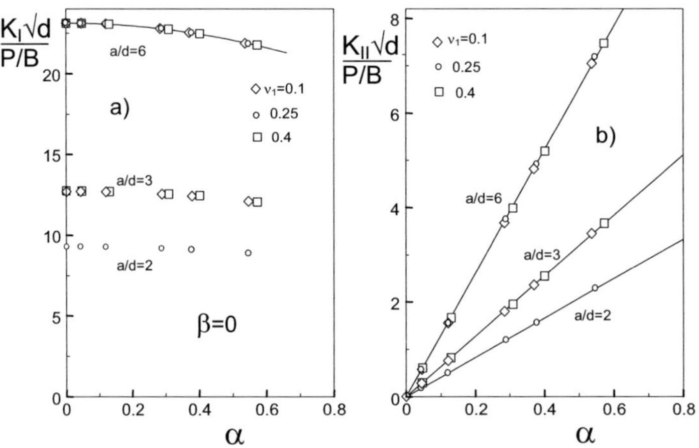

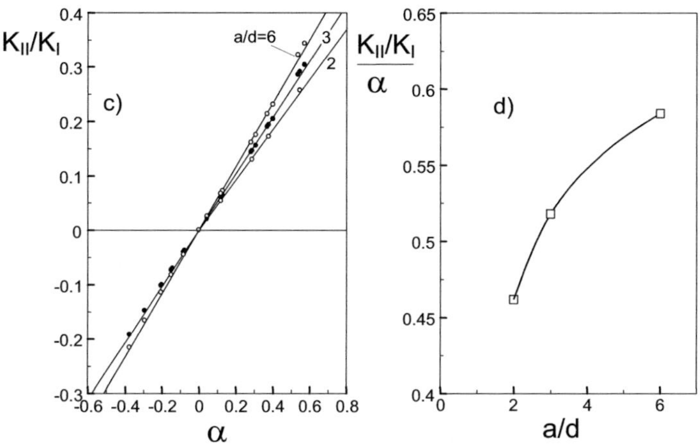

Fig. E2.11 Stress intensity factors K_I (a) and K_{II} (b) as functions of the first Dundurs parameter α for different a/d (β=0), c) and d): mode-mixity K_{II}/K_I.

The dependencies on α can be approximated as

$$\frac{K_I(\alpha)}{K_I(0)} \cong 1 - 0.171\alpha^2 \qquad (E2.2.5a)$$

and

$$K_{II}(\alpha) \propto \alpha \qquad (E2.2.5b)$$

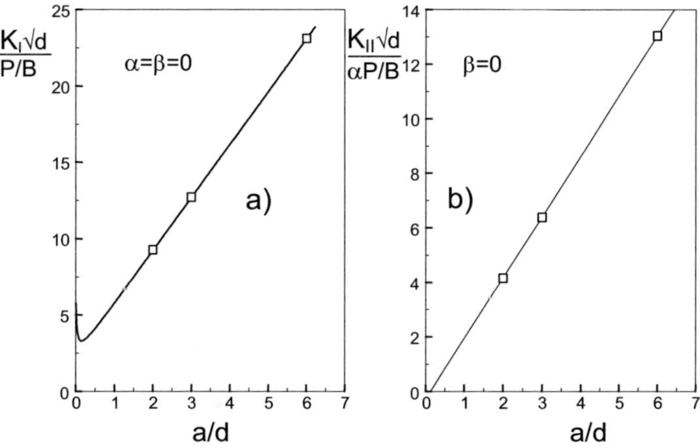

Fig. E2.12 a) Stress intensity factor of a homogeneous material versus a/d (curve given by eq.(E2.2.7a)), b) K_{II}/α for several ratios a/d.

Figure E2.11c represents the "mode mixity" K_{II}/K_I. Also this quantity is found to be linearly dependent on α. The coefficient of proportionality is entered in Fig. E2.11d. For the special case of a homogeneous material ($\alpha=\beta=0$), the mode-I stress intensity factor can be determined from the weight function which reads [E2.8]

$$h = \sqrt{\frac{12}{d}}\left(\frac{a-\xi}{d}+\lambda\right)+\sqrt{\frac{2}{\pi(a-\xi)}}\,\exp\left(-\sqrt{12\frac{a-\xi}{d}}\right) \qquad (E2.2.6)$$

with $\lambda=0.68$. The stress intensity factor for loading at $\xi=0$ then results as

$$\frac{K_I}{P/B} = \sqrt{\frac{12}{d}}\left(\frac{a}{d}+\lambda\right)+\sqrt{\frac{2}{\pi a}}\,\exp\left(-\sqrt{12\frac{a}{d}}\right) \qquad (E2.2.7a)$$

This solution is introduced in Fig. E2.12a as the curve. Equation (E2.2.7a) may be simplified for $a/d>1$ by

$$K_I \cong \frac{P}{B}\sqrt{\frac{12}{d}}\left(\frac{a}{d}+\lambda\right) \qquad (E2.2.7b)$$

Together with eq.(E2.2.5a), the following expression is obtained for the case of dissimilar materials

$$K_I \cong \frac{P}{B}\sqrt{\frac{12}{d}}\left(\frac{a}{d}+\lambda\right)(1-0.171\alpha^2) \qquad (E2.2.8)$$

Figure E2.12b displays the mode-II stress intensity factor as a function of a/d. The straight line dependency (solid line) is represented by

$$K_{II} \cong \frac{P}{B\sqrt{d}}\frac{20}{9}\left(\frac{a}{d}-\frac{1}{8}\right)\alpha \qquad (E2.2.9)$$

The constant stress terms σ_{01} and σ_{02} are represented in Fig. E2.13. These results may be approximated by the straight line relations of

$$\sigma_{01} = T(1-\lambda\alpha) \quad , \quad 0\leq\alpha\leq 0.5 \qquad (E2.2.10)$$

$$\sigma_{02} = T(1+0.8\alpha) \qquad (E2.2.11)$$

where T is the constant stress term at $\alpha=0$, i.e. the T-stress of the homogeneous specimen.

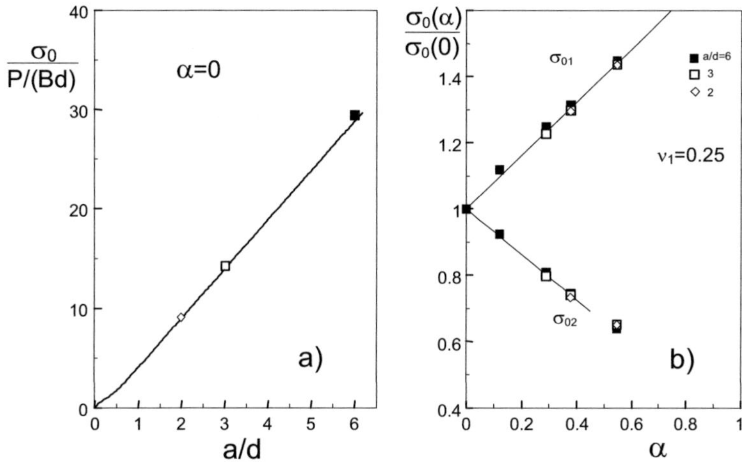

Fig. E2.13 a) Constant stress term of the homogeneous material versus *a/d* (curve given by eq.(E2.2.13)), b) plot of the constant stress terms normalised to the constant stress of a homogeneous material for several ratios of *a/d* (β=0).

The T-stress caused by a point force at the distance ξ from the end of the bar is given as [E2.9]

$$\frac{T}{P/(Bd)} = \lambda\left(\tfrac{9}{8}\sqrt{\frac{a-\xi}{d}} + \tfrac{5}{2}\left(\frac{a-\xi}{d}\right)^{3/2}\right) + (1-\lambda)\left(4.95\frac{a-\xi}{d} - 0.9\right) \qquad \text{(E2.2.12)}$$

with an interpolation function

$$\lambda = \exp\left(-2\left(\frac{a-\xi}{d}\right)^2\right) \qquad \text{(E2.2.13)}$$

and for the special case ξ=0, it results

$$\frac{T}{P/(Bd)} = 4.95\frac{a}{d} - 0.9 + \exp\left(-2\left(\frac{a}{d}\right)^2\right)\left(0.9 + \tfrac{9}{8}\sqrt{\frac{a}{d}} + \tfrac{5}{2}\left(\frac{a}{d}\right)^{3/2} - 4.95\frac{a}{d}\right) \qquad \text{(E2.2.14)}$$

Equation (E2.2.14) is introduced in Fig. E2.13a as the solid curve.

E2.3 The compact tension (CT) specimen

The standard geometry of the compact tension (CT) specimen is illustrated in Fig. E2.14. The thickness again is B. Computations similar to those given in detail for the DCB specimen were performed for the CT specimen. The graphical representations concentrate on the most essential results only.

Figure E2.15 shows the mode-I stress intensity factor solution. The circles correspond to the stress intensity factor K_I and the squares represent the energy release rate expressed by eq.(E2.2.1) in terms of the effective stress intensity factor K_{eff}. Figure E2.15b gives the data for $a/W{=}0.5$ in higher resolution.

From Fig. E2.15, it becomes obvious that

-    the dependence on α is negligible and

-    the effective stress intensity factor is nearly identical with the mode-I contribution.

This behaviour is due to the very small mode-II stress intensity factor contributions which are less than 10% of the mode-I stress intensity factors, as can be seen from Fig. E2.16a. As evident from this diagram, the dependency of K_{II} on the Dundurs parameter α is linear. Figure E2.16b represents the mode mixity K_{II}/K_I. Also this plot reflects the minor influence of K_{II} for the CT specimen, as obvious from comparing with the mode mixity for the DCB pecimen (Fig. E2.11c). Figure E2.16c finally shows the slope of the straight lines in Fig. E2.16b.

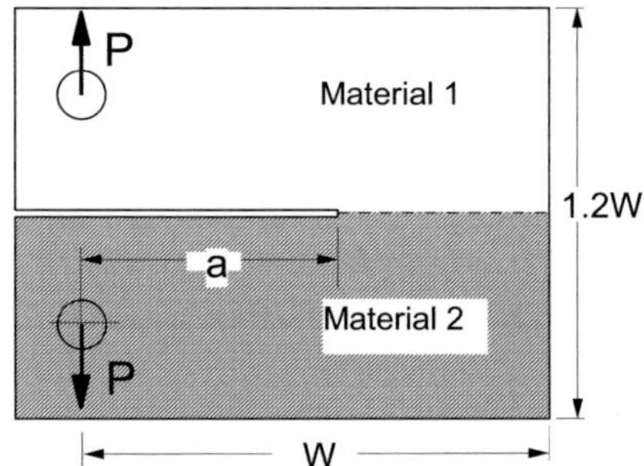

Fig. E2.14 The compact tension test specimen.

In Fig. E2.17 the stress intensity factor K_I of the homogeneous material ($\alpha{=}\beta{=}0$) is plotted versus the relative crack length a/W as the squares. The well-known stress intensity factor solution for the CT specimen made of homogeneous material is [E2.10]

$$K_{\mathrm{I}} = \frac{P}{B\sqrt{W}} \, \frac{(2+\eta)(0.886 + 4.64\eta - 13.32\eta^2 + 14.72\eta^3 - 5.6\eta^4)}{(1-\eta)^{3/2}}, \quad \eta = a/W \quad (E2.3.1)$$

This dependency is represented as the curve in Fig. E2.17a. Good agreement is obvious.

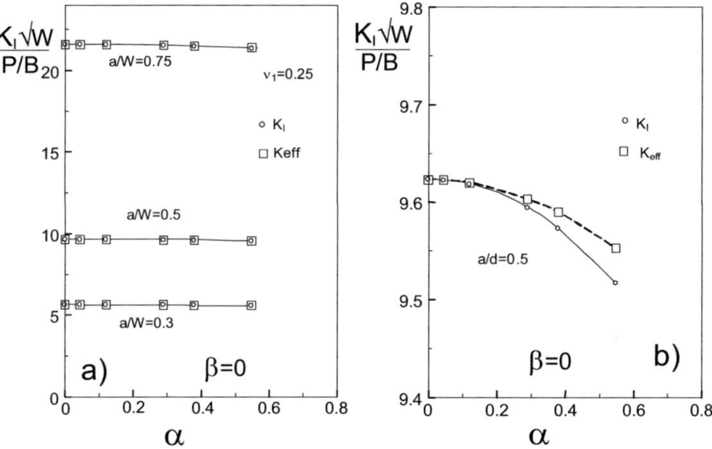

Fig. E2.15 Stress intensity factor K_{I} and effective stress intensity factor K_{eff} (representing the energy release rate).

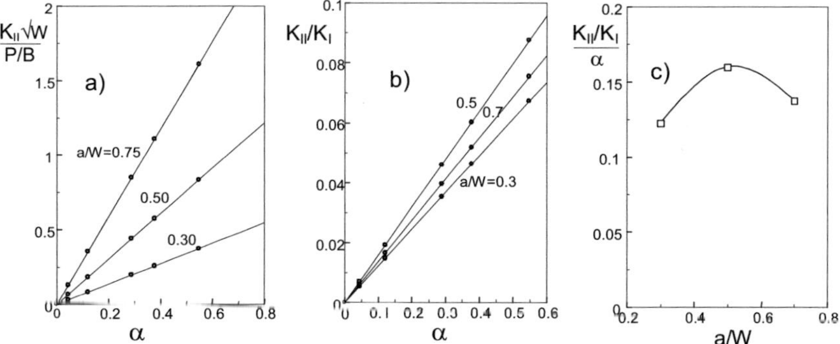

Fig. E2.16 a) Stress intensity factor K_{II}, b) mode mixity $K_{\mathrm{II}}/K_{\mathrm{I}}$, and c) slope of the mode mixity straight lines in b).

The two constant x-stress terms σ_{01} and σ_{02} are given in Fig. E2.18a, normalised on the stress term at $\alpha=0$. The x-stress of the homogeneous specimen is identical with the

T-stress and plotted in Fig. E2.18b versus the relative crack size a/W. At $a/W<0.6$ and $\alpha<0.6$, the following relations are proposed:

$$\sigma_{01} = T(1+0.8\,\alpha) \qquad\qquad (E2.3.2)$$

$$\sigma_{02} = T(1-0.85\,\alpha) \qquad\qquad (E2.3.3)$$

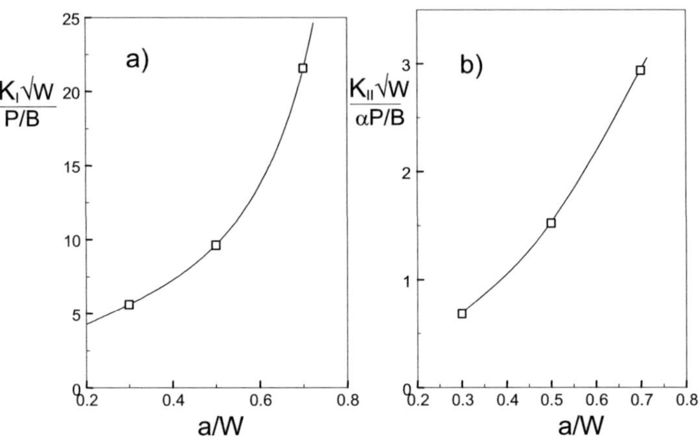

Fig. E2.17 a) Mode-I stress intensity factor K_{I} of the homogeneous material ($\alpha=\beta=0$), symbols: FE results, curve: eq.(E2.3.1) proposed by Srawley [E2.10], b) slope of the straight lines in Fig. E2.16a.

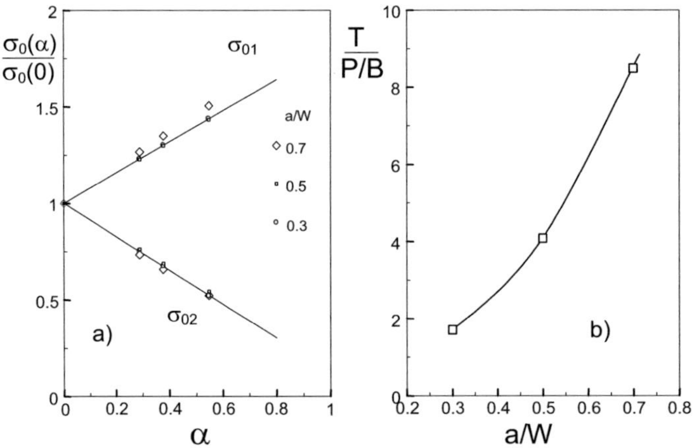

Fig. E2.18 a) Constant stress terms normalised to the constant stress of homogeneous material at variable ratios a/W ($\beta=0$), straight lines: eqs.(E2.3.2) and (E2.3.3), b) constant stress term (T-stress) of homogeneous material versus a/W.

E2.4 DCDC test specimen

The "double cleavage drilled compression" (DCDC) specimen shown in Fig. E2.19 is used to determine stable and subcritical crack growth under mixed-mode loading conditions (e.g. [E2.11, E2.12, E2.13]).

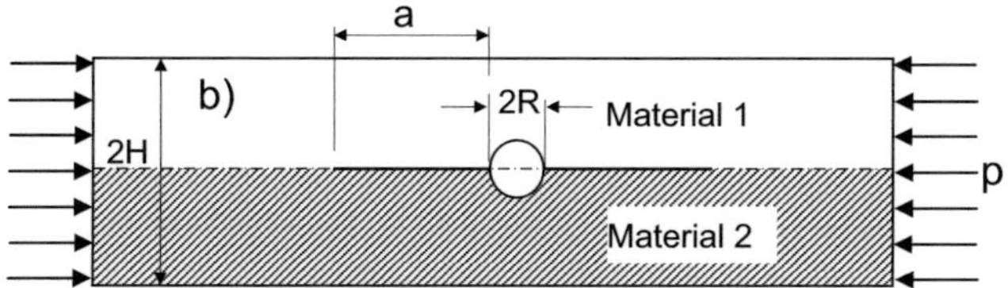

Fig. E2.19 DCDC specimen with central hole

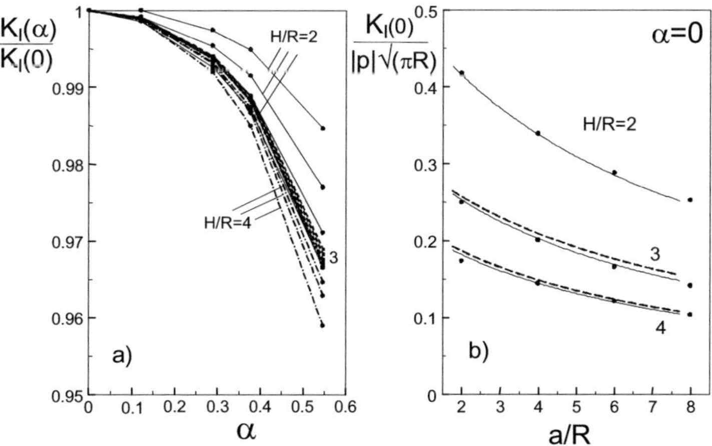

Fig. E2.20 a) Influence of Dundurs parameter α on K_I, b) stress intensity factor K_I of homogeneous material, circles: FE-results, dashed curves: eq.(E2.4.2) [E2.11], solid curves: eq.(E2.4.3) [E2.14].

Figure E2.20a shows the mode-I stress intensity factor K_I normalised to the value of homogeneous material (i.e. for $\alpha=\beta=0$) versus α. The plots for $H/R=3$ and 4 may be expressed by the common relation of

$$K_1 = K_1(0)\,(1 - 0.107\,\alpha^2)\qquad\text{(E2.4.1)}$$

For the case of a homogeneous material ($\alpha=\beta=0$), the mode-I stress intensity factor at $a/R \geq 4$ was given by He et al. [E2.11] as

$$\frac{|p|\sqrt{\pi R}}{K_I(0)} = \frac{H}{R} + \left[0.235\frac{H}{R} - 0.259\right]\frac{a}{R} \qquad (E2.4.2)$$

This solution is introduced as the dashed curves in Fig. E2.20b. A solution proposed in [E2.14] reads

$$\frac{|p|\sqrt{\pi R}}{K_I(0)} = 1.1163\frac{H}{R} - 0.3703 + \left[0.216\frac{H}{R} - 0.1575\right]\frac{a}{R} \qquad (E2.4.3)$$

which is represented by the solid curves in Fig. E2.20b. Equation (E2.4.1) in combination with (E2.4.2) or (E2.4.3) allows to compute the mode-I stress intensity factor for the DCDC specimen made of dissimilar materials.

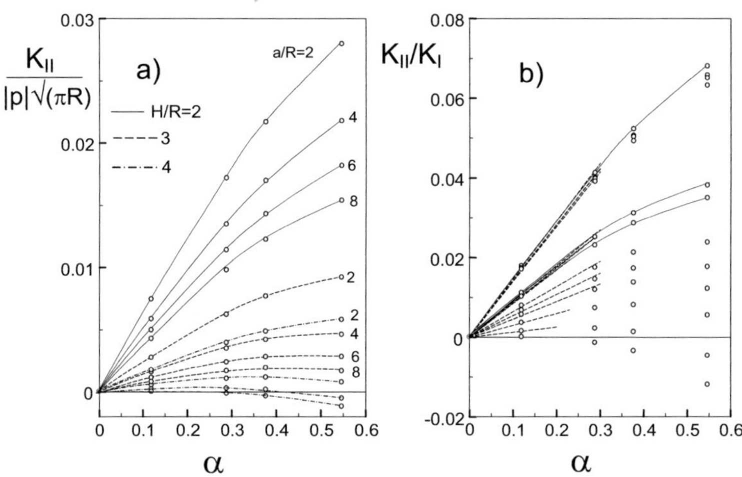

Fig. E2.21 a) Mode-II stress intensity factor and b) mode mixity K_{II}/K_I versus Dundurs parameter α.

The mode-II stress intensity factor K_{II} is plotted in Fig. E2.21a. Figure E2.21b represents the mode mixity $K_{II}/K_I=f(\alpha)$. If $\alpha<0.3$, mode mixity may be approximated by linear relations as

$$K_{II}/K_I = C\alpha \qquad (E2.4.4)$$

with the coefficient C compiled in Table E2.1.

In Fig. E2.22 the constant stress terms are plotted for $a/R=2$. At $H/R>3$ and $\alpha<0.6$, the constant stresses may be estimated roughly by

$$\sigma_{01}/T \approx 1 + 0.6678\,\alpha + 2.4987\,\alpha^2 \qquad (E2.4.5)$$

$$\sigma_{02}/T \approx 1 - 0.9329\,\alpha + 0.4008\,\alpha^2 \qquad (E2.4.6)$$

These dependencies are presented as the solid curves in Fig. E2.22b. The T-stress T of the homogeneous material can be expressed by [E2.14]

$$T/|p| = \frac{1}{1.11\,H/R - 1.157 + (0.213\,H/R - 0.283)\,a/R} - 1 \qquad (E2.4.7)$$

H/R	$a/R=2$	4	6	8
2	0.145	0.141	0.140	0.138
3	0.0494	0.0630	0.0531	0.0441
4	0.0842	0.0298	0.0126	0

Table E2.1 Coefficients C for mode mixity according to eq.(E2.4.4), $\alpha<0.3$.

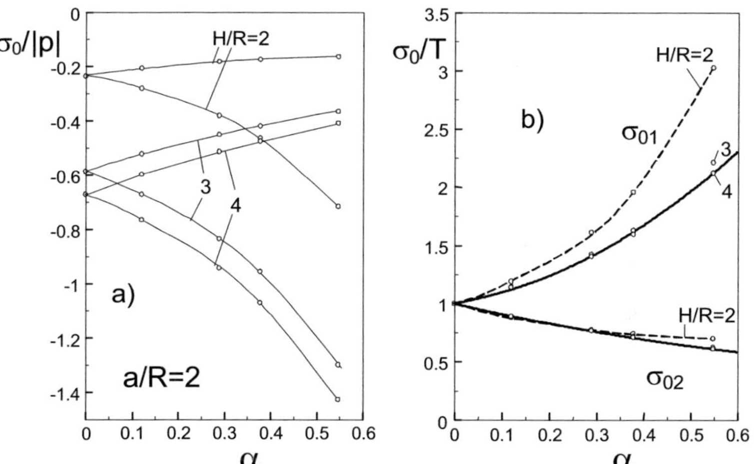

Fig. E2.22 a) Constant stress terms of the DCDC test specimen, b) normalised to the T-stress.

E2.5 Bending bar

The bending bar made of dissimilar materials was studied very early. A large number of references is given in [E2.15]. In most papers, the energy release rate was considered as the driving force in fracture mechanics tests. Therefore, the present focus is on the constant stress terms. Figure E2.23 shows the geometrical data.

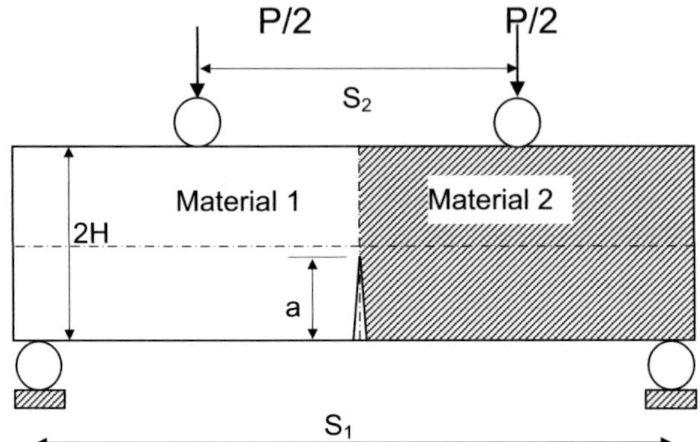

Fig. E2.23 4-point bending specimen with a crack at the interface.

For the homogeneous test specimen, the stress intensity factor and the T-stress are well known. The stress intensity factor K_I is

$$K_I = \sigma_b \sqrt{\pi a}\, F \qquad (E2.5.1)$$

with the bending stress σ_b

$$\sigma_b = \frac{3P(S_1 - S_2)}{2BW^2} \qquad (E2.5.2)$$

and [E2.8]

$$F = \frac{1.1215}{(1-\eta)^{3/2}} \left[\frac{5}{8} - \frac{5}{12}\eta + \frac{1}{8}\eta^2 + 5\eta^2(1-\eta)^6 + \frac{3}{8}\exp(-6.1342\,\eta/(1-\eta)) \right] \qquad (E2.5.3)$$

with $\eta = a/W$. This relation is plotted as the curve in Fig. E2.24a. The squares represent numerical solutions obtained by the finite element computations. Good agreement can be seen. The related T-stress can be expressed by

$$\frac{T}{\sigma_b} = \frac{-0.526 + 2.481\eta - 3.553\eta^2 + 2.6384\eta^3 - 0.9276\eta^4}{(1-\eta)^2} \qquad (E2.5.4)$$

This dependency is plotted as the curve in Fig. E2.24b. Also in this case, the squares result from FE computations. Best agreement with eq.(E2.5.4) is evident. In Fig. E2.25a, the mode-I stress intensity factor of the specimen made of dissimilar materials is normalised to the stress intensity factor according to eqs.(E2.5.1-E2.5.3) and plotted versus the Dundurs parameter α. The influence of α is negligible at $\alpha<0.5$. Figure E2.25b shows the mixed-mode ratio K_{II}/K_I for several relative crack lengths a/W. In Fig. E2.25c, the steepness of the curves K_{II}/K_I vs. α is shown, defining the coefficient λ_1 in

$$K_{II}/K_I = \lambda_1 \alpha \ . \tag{E2.5.5}$$

In Fig. E2.25d, the ratio between the effective stress intensity factor K_{eff} (representing the energy release rate via eq.(E2.1.6)) and the mode-I contribution K_I is plotted. Maximum deviations of less than 0.4% are visible. From this result, it can be concluded that the solution for homogeneous material can be applied to the computation of energy release rates.

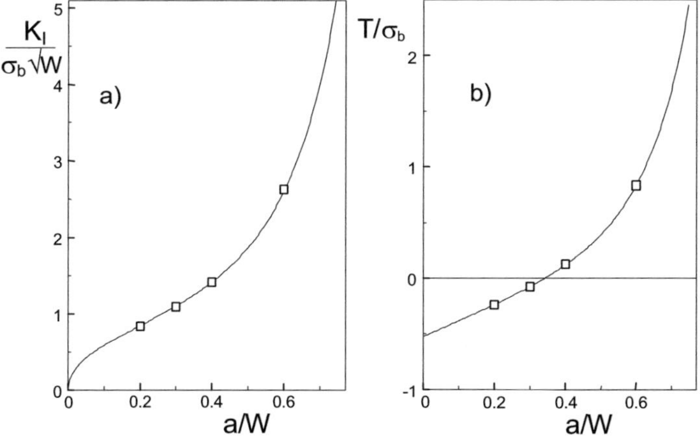

Fig. E2.24 a) Mode-I stress intensity factor K_I of homogeneous material ($\alpha=\beta=0$), symbols: FE results, curve: eq.(E2.5.3), b) T-stress solution eq.(E2.5.4).

The constant stress terms for three crack lengths are plotted in Fig. E2.26a, resulting in

$$\sigma_{01} = T + \lambda_2 \alpha \tag{E2.5.6}$$

$$\sigma_{02} = T - \lambda_2 \alpha \tag{E2.5.7}$$

with T given by eq.(E2.5.4).

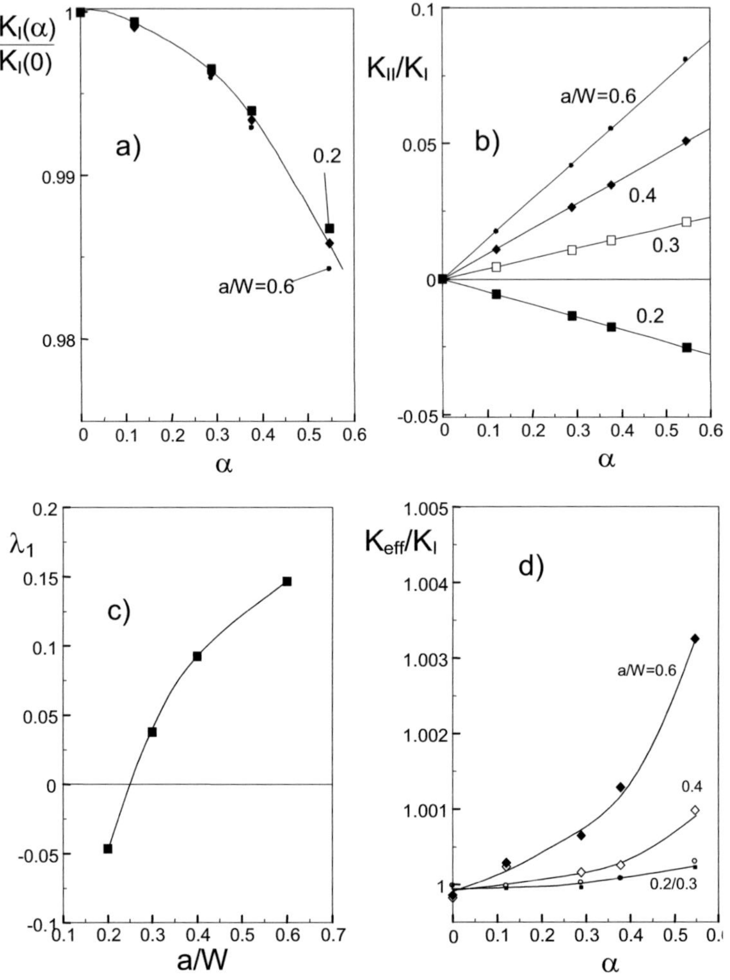

Fig. E2.25 a) Mode-I stress intensity factor K_I of dissimilar materials, b) mixed-mode ratio, c) slope of the curves in (b), defining the coefficient λ_1 in eq.(E2.5.5), d) energy release rate expressed by the effective stress intensity factor according to eq.(E2.1.6).

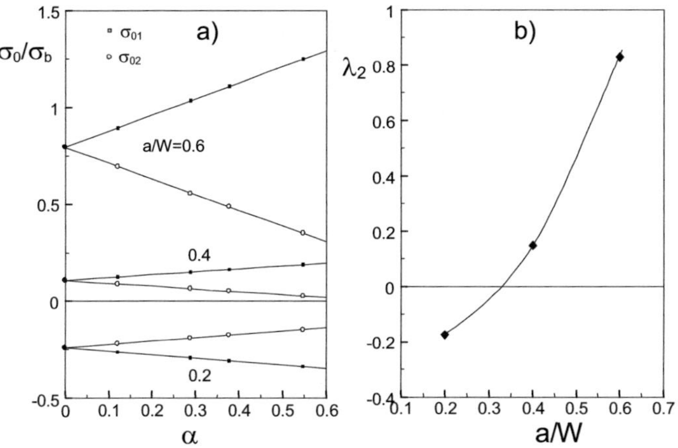

Fig. E2.26 a) Constant stress terms normalised to the bending stress, b) coefficient λ_2 for eqs.(E2.5.6) and (E2.5.7), representing the slopes in (a).

E2.6 Opposite roller test

An experimental set-up for a fracture mechanics test with completely stable crack propagation as developed in [E2.16] for homogeneous materials is shown in Fig. E2.27. A pre-notched bar is loaded via four opposite rollers. The effect of dissimilar materials will be studied below.

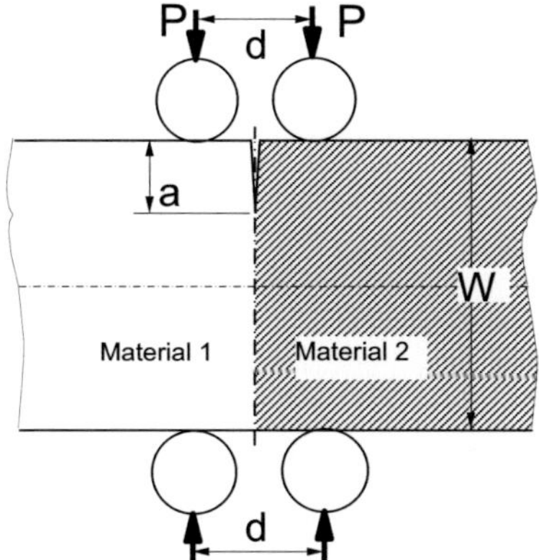

Fig. E2.27 Controlled fracture test device with load application via four symmetrical rollers.

Figure E2.28 represents the mode-I stress intensity factor K_I as a function of the Dundurs parameter α and the relative crack length a/W. From the plot in Fig. E2.28b, it is clearly visible that the influence of crack length on the normalised stress intensity factor $K_I(\alpha)/K_I(0)$ is negligible. The value of $K_I(0)$ is identical with the stress intensity factor solution for homogeneous material. The solution obtained by the weight function technique reads [E2.16]

$$K_I = \frac{2P}{B\sqrt{W}}(0.905\eta^{1/2} - 3.358\eta^{3/2} + 3.857\eta^{5/2} + 1.4425\eta^{7/2} - 3.873\eta^{9/2}) \quad (E2.6.1)$$

with $\eta = a/W$. Equation (E2.6.1) is plotted in Fig. E2.29 together with the data obtained from a FE analysis. Good agreement is obvious.

Figure E2.30 shows the mode-II stress intensity factor as a function of the Dundurs parameter α and the relative crack length a/W. The linear dependencies shown in Fig. E2.30a can be expressed as

$$K_{II} = C_{II}\frac{P}{B\sqrt{W}}\alpha \quad (E2.6.2)$$

with the coefficient C_{II} plotted in Fig. E2.30b versus the relative crack length. Finally, Fig. E2.30c illustrates the mixed-mode ratio K_{II}/K_I. Since the mode-II stress intensity factor is small compared to the mode-I stress intensity factor at $a/W=0.2$ and 0.4, it is self-evident that the effective stress intensity factor K_{eff} representing the energy release rate cannot differ significantly from the mode-I value K_I. Due to this fact, a separate plot of the effective stress intensity factor did not appear to be necessary.

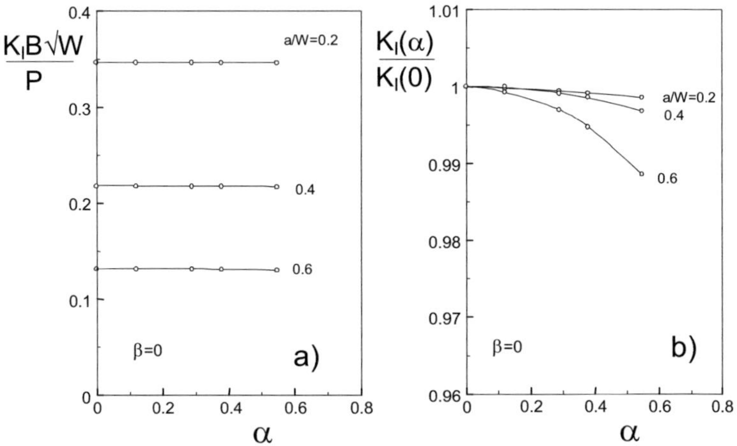

Fig. E2.28 a) Influence of Dundurs parameter α on K_I, b) normalised representation of a).

The constant stress terms are given in Fig. E2.31a. The straight line behaviour may be expressed by the relations

$$\sigma_{01} = T(1 + C_1\alpha) \tag{E2.6.3}$$

$$\sigma_{02} = T(1 - C_2\alpha) \tag{E2.6.4}$$

The T-stress data obtained with FE are shown by the symbols in Fig. E2.32. A solution tabulated in [E2.17] and interpolated using cubic splines is entered as the solid curve. Also in this case, good agreement is visible.

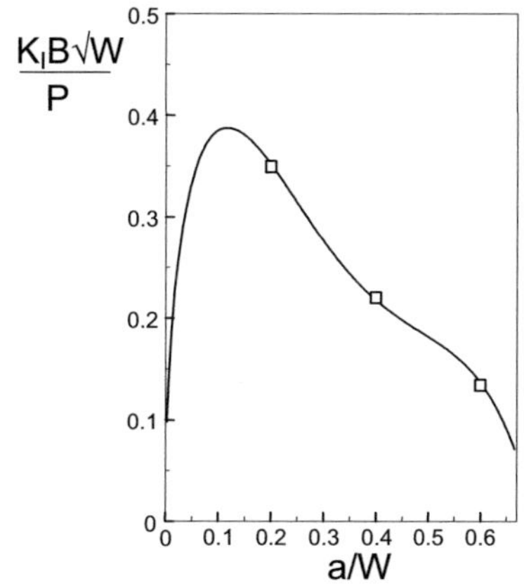

Fig. E2.29 Stress intensity factor solution for homogeneous material; curve: eq.(E2.6.1) [E2.17], squares: FE results.

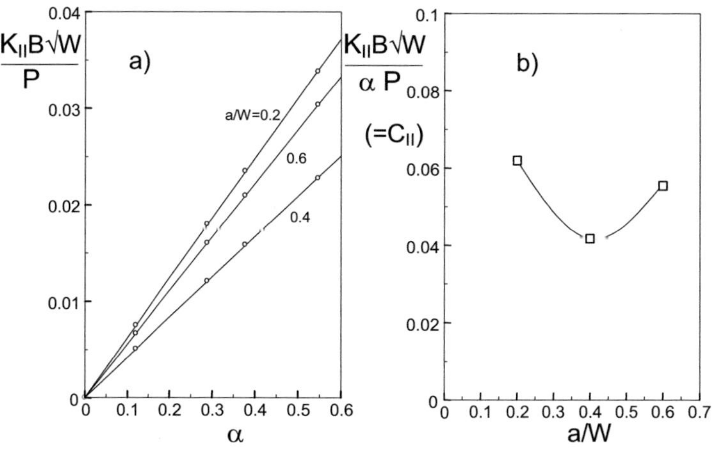

29

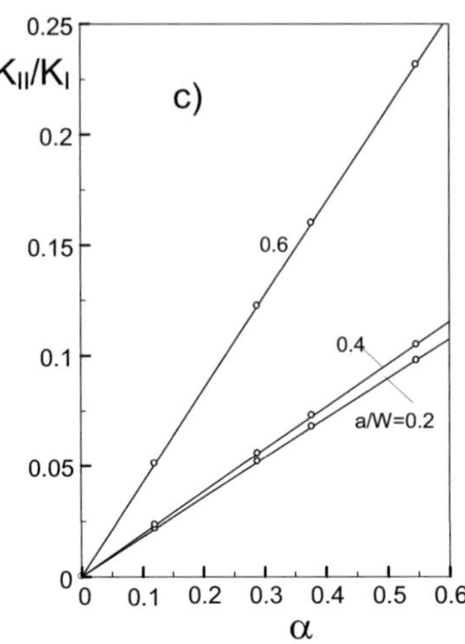

Fig. E2.30 a) Influence of Dundurs parameter α on K_{II}, b) steepness of the curves of a), c) mixed-mode ratio K_{II}/K_I.

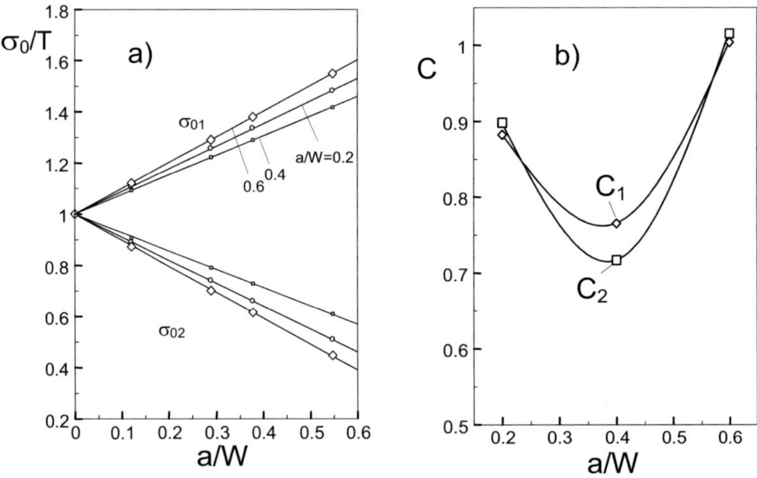

Fig. E2.31 a) Constant stress terms normalised to the T-stress, b) coefficients for eqs. (E2.6.3) and (E2.6.4).

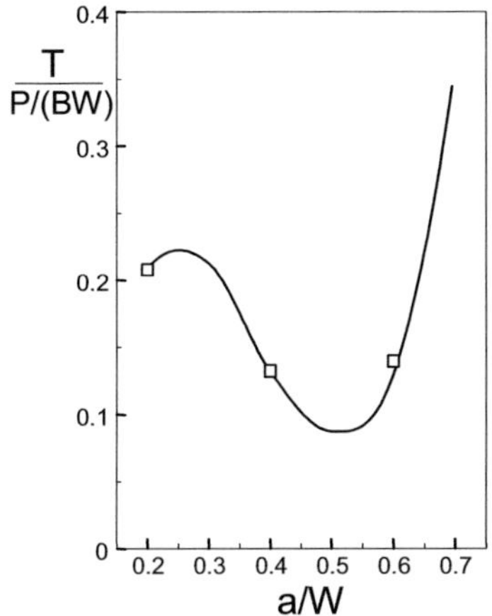

Fig. E2.32 T-stress from the FE analysis (symbols) compared with a solution tabulated in [E2.16] and interpolated with cubic splines.

E2.7 Cracks in bending bars normal to an interface

E2.7.1 Stresses in a bending bar

Figure E2.33 shows a bending bar made of dissimilar materials. The crack is normal to the interface. It is clear that in case of the crack-tip located directly at the interface the description of the loading parameter by conventional stress intensity factors is no longer possible (at least in the general case of $\beta\neq0$). Nevertheless, stress intensity factor K_I exists if the crack tip is located in one of the two materials.

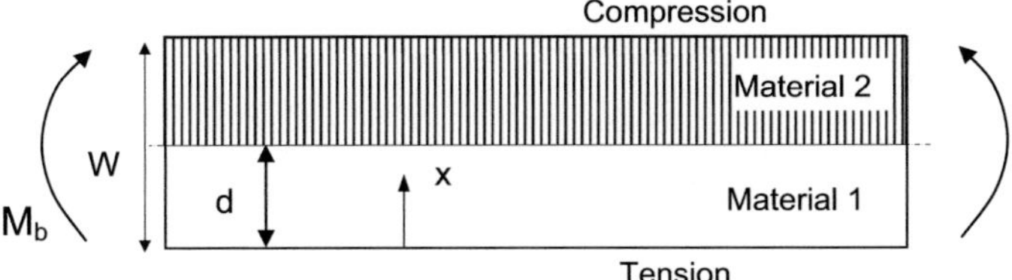

Fig. E2.33 Bending bar made of dissimilar materials.

For the special case of

$$d=W/2$$

(considered in the following) an applied bending moment M_b generates the outer fibre bending stress σ_t at the tensile and σ_c at the compressive surface

$$\sigma_{t,c} = \sigma_0 \begin{cases} \dfrac{4R(3+R)}{1+14R+R^2} & \textit{for tension} \\[3mm] -\dfrac{4+12R}{1+14R+R^2} & \textit{for compression} \end{cases} \qquad \text{(E2.7.1)}$$

with the modulus ratio $R=E_1/E_2$ and the *formally* computed bending stress

$$\sigma_0 = \frac{M_b}{\frac{1}{6}BW^2} \qquad \text{(E2.7.2)}$$

The strain distribution through the bar is in absence of the crack

$$\varepsilon = \frac{\sigma_0}{E_1}\frac{4R(3+R)-16R(1+R)x/W}{1+14R+R^2} \qquad \text{(E2.7.3)}$$

with the neutral axis ($\varepsilon=0$) at

$$\frac{x_0}{W} = \frac{3+R}{4(1+R)} \qquad \text{(E2.7.4)}$$

The strain distributions in the uncracked bar are shown in Fig. E2.34.

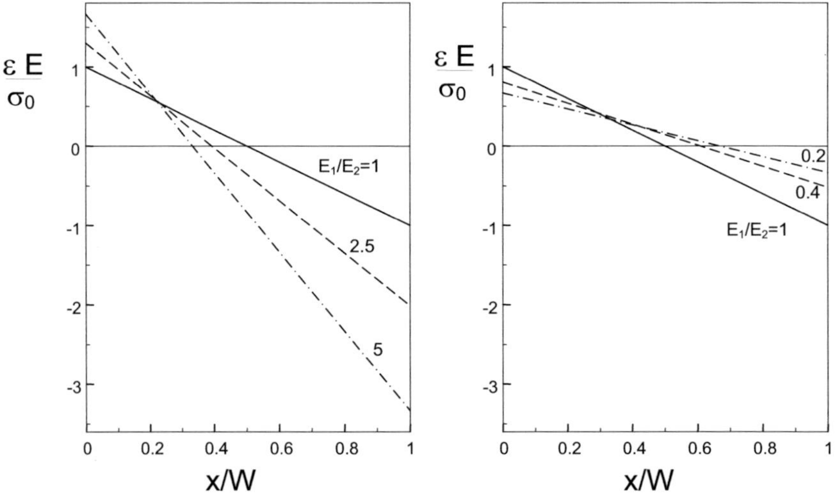

Fig. E2.34 Strain distributions through a bending bar of dissimilar materials.

The stress distributions described by

$$\sigma = \sigma_0 \begin{cases} \dfrac{4R(3+R)-16R(1+R)x/W}{1+14R+R^2} & \text{for } x/W < 0.5 \\[3mm] \dfrac{4(3+R)-16(1+R)x/W}{1+14R+R^2} & \text{for } x/W \geq 0.5 \end{cases} \qquad (E2.7.5)$$

are shown in Fig. E2.35. For $E_1/E_2 > 1$, the outer fibre tensile stress is increased with respect to a homogeneous material. A reduction of the maximum tensile stress follows in the case $E_1/E_2 < 1$. Stress distributions for arbitrary ratios of d/W are given by Noda et al. [E2.18].

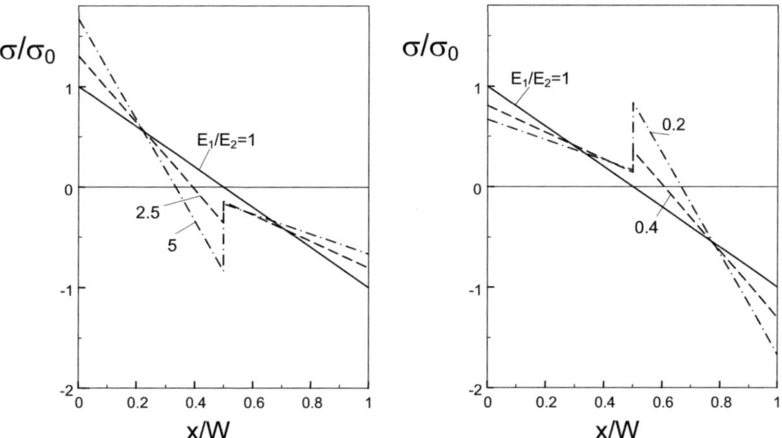

Fig. E2.35 Stress distributions through a bending bar of dissimilar materials.

E2.7.2 Stress intensity factors for 3-point bending

Stress intensity factor solutions for cracks approaching and penetrating an interface are known from literature. Many results are available in literature (see e.g. [E2.19]) predominantly for infinite and semi-infinite bodies. Edge-cracked surface layers were very early studied by Lu and Erdogan [E2.20], Fujino et al. [E2.21], Rizk and Erdogan [E2.22], and approximate weight functions were derived in [E2.23]. Noda et al. [E2.18] considered finite edge-cracked bars under tension and bending. In the latter case especially the ratios $E_1/E_2=3$ and 1/3 were chosen. For special applications to the dental materials Enamel and Dentin, the modulus ratios $E_1/E_2=2.5$ and 0.4 are of interest. Strength tests on these materials are commonly performed in 3-point bending tests with rather short relative supporting span S/W. For understanding the crack path in such specimens also the T-stresses and biaxiality ratios are necessary.

Figure E2.36 shows a 3-point bending test with a crack in the bar. The crack is normal to the interface. It is clear that in case of the crack-tip located directly at the interface the description of the loading parameter by conventional stress intensity factors is no longer possible (at least in the general case of $\beta\neq0$) [E2.24]. Nevertheless, stress intensity factor K_I exist if the crack tip is located in one of the two materials.

Very often the material properties do not change abruptly at an interface but vary more smoothly in a finite layer. In such cases the fracture mechanics analysis is again based on stress intensity factors. Numerous studies exist in this field ([E2.25], [E2.26]) dealing with functionally graded materials ([E2.27], [E2.28]) and thermal loading problems ([E2.29], [E2.30]).

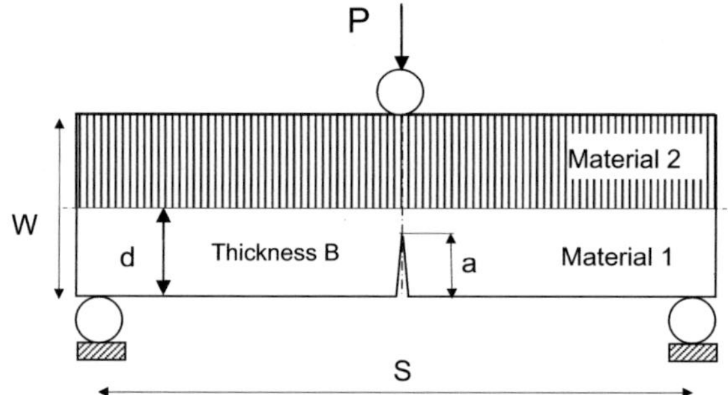

Fig. E2.36 3-point bending specimen with a crack normal to an interface.

a/W	$E_1/E_2{=}1$	1/0.75	1/0.4	1/0.2
0	1.10	1.185	1.432	1.833
0.2	1.001	1.095	1.358	1.788
0.4	1.182	1.331	1.708	2.291
0.48	1.355	1.623	2.302	3.167
0.52	1.472	1.518	1.607	1.685
0.6	1.811	1.820	1.834	1.847

Table E2.2 Geometric function according to eq.(E2.7.6) for 3-point bending ($E_1/E_2{\geq}1$, $S/W{=}5$).

Results of the geometric functions will be reported by the geometric function F defined by

$$K_I = \sigma_0 F \sqrt{\pi a} \qquad \text{(E2.7.6)}$$

The geometric function for a crack with $a/W\rightarrow0$ is given by

$$F = 1.1\frac{4R(3+R)}{1+14R+R^2} \qquad (E2.7.7)$$

Finite element results for $S=5\,W$, the range of $0.2 \leq E_1/E_2 \leq 5$ (Poisson's ratio 0.25) are potted in Fig. E2.37 and compiled in Table E2.2 for interpolations. For the cases $E_1/E_2 \leq 3$ and 1/3 in pure bending see Noda et al. [E2.18]. The curves approaching $a/W=0.5$ were tentatively plotted for $0.48 \leq a/W \leq 0.52$.

A finite element study was performed for several ratios of E_1/E_2 with $v_1=v_2$ For the computations under plane strain conditions the special geometry of $S/W=5$ was considered. Data for other ratios of S/W close to this value may be estimated by

$$F_{(S/W;R)} \cong \frac{F_{(5;R)}}{F_{(5;1)}} F_{(S/W;1)} \qquad (E2.7.8)$$

a/W	$E_1/E_2=1$	0.75	0.4	0.2
0	1.10	1.026	0.885	0.733
0.2	1.001	0.918	0.809	0.583
0.4	1.182	1.044	0.769	0.514
0.48	1.355	1.115	0.700	0.400
0.52	1.472	1.423	1.307	1.180
0.6	1.811	1.803	1.778	1.748

Table E2.3 Geometric function according to eq.(E2.7.6) for 3-point bending ($E_1/E_2 \leq 1$, $S/W=5$).

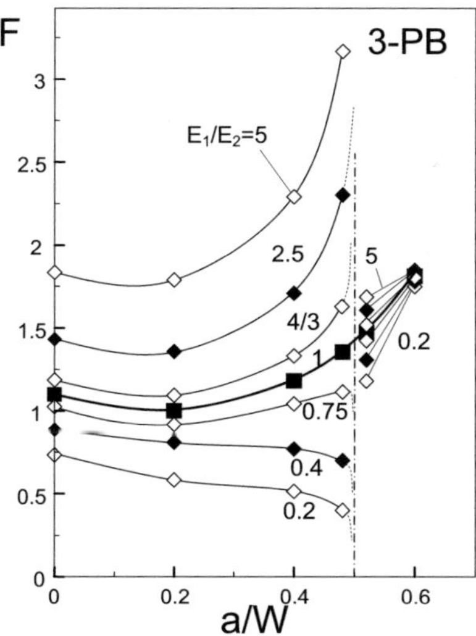

Fig. E2.37 Geometric functions for stress intensity factor in 3-point bending at different ratios E_1/E_2.

E2.7.3 Stress intensity factors for 4-point bending

Figure E2.38 shows the 4-point bending test for $S_2/W{=}2.5$ studied by FE. The stress intensity factors are again defined by eq.(E2.7.6) and (E2.7.2). The results are represented in Fig. E2.39. Also Tables E2.4 and E2.5 compile these data.

In 4-point bending it holds for $a/W{\rightarrow}0$

$$F = 1.1215\frac{4R(3+R)}{1+14R+R^2} \qquad (E2.7.9)$$

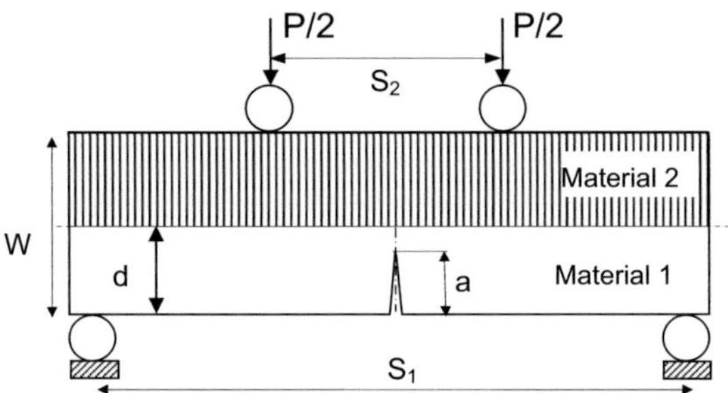

Fig. E2.38 4-point bending specimen with a crack normal to an interface.

a/W	$E_1/E_2{=}1$	1/0.75	1/0.4	1/0.2
0	1.122	1.208	1.460	1.869
0.2	1.055	1.141	1.382	1.783
0.4	1.260	1.410	1.792	2.381
0.48	1.439	1.725	2.451	3.371
0.52	1.562	1.605	1.688	1.762
0.6	1.914	1.922	1.937	1.948

Table E2.4 Geometric function according to eq.(E2.7.6) for 4-point bending, $S_2/W{=}2.5$.

a/W	$E_1/E_2{=}1$	0.75	0.4	0.2
0	1.122	1.046	0.903	0.748
0.2	1.055	0.978	0.826	0.657
0.4	1.260	1.119	0.836	0.571
0.48	1.439	1.185	0.749	0.435
0.52	1.562	1.516	1.402	1.265
0.6	1.914	1.904	1.880	1.855

Table E2.5 Geometric function according to eq.(E2.7.6) for 4-point bending, $S_2/W{=}2.5$.

Stress intensity factors for other ratios of $S_2/W \neq 2.5$ can be estimated by using a reation similar to eq.(E2.7.8).

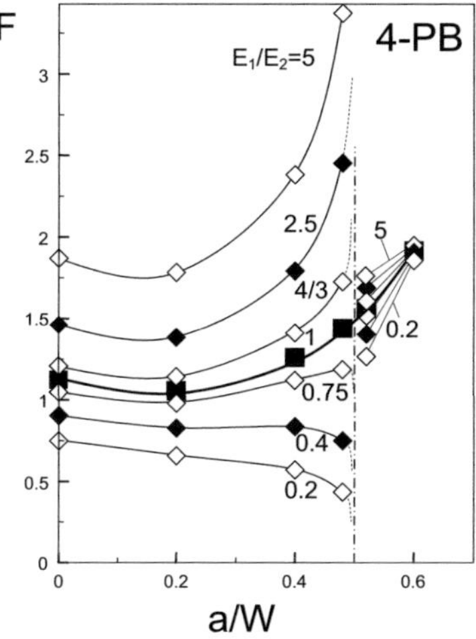

Fig. E2.39 Geometric functions for stress intensity factor in 4-point bending at different ratios E_1/E_2, S_2/W=2.5.

E2.7.4 Biaxiality ratio

The biaxiality ratio β by Leevers and Radon [E2.31] as the dimensionless representation of T

$$\beta = \frac{T\sqrt{\pi a}}{K_1} = \frac{T}{\sigma_0 F} \tag{E2.7.10}$$

is compiled in Tables E2.6 and E2.7 for 3-point bending and in E2.8 and E2.9 for 4-point bending. Figure E2.39 represents the data for 3-point bending.

a/W	E_1/E_2=1	1/0.75	1/0.4	1/0.2
0	-0.469	-0.469	-0.469	-0.469
0.2	-0.202	-0.221	-0.157	-0.101
0.4	0.001	-0.011	0.016	0.133
0.48	0.097	-0.163	-0.585	-0.783
0.52	0.149	0.071	-0.159	-0.459
0.6	0.253	0.266	0.279	0.272

Table E2.6 Biaxiality ratio according to eq.(E2.7.10) for 3-point bending, S/W=5.

a/W	E_1/E_2=1	0.75	0.4	0.2
0	-0.469	-0.469	-0.469	-0.469
0.2	-0.202	-0.224	-0.199	-0.124
0.4	0.001	0.025	0.101	0.224
0.48	0.097	0.396	1.149	2.031
0.52	0.149	0.183	0.154	-0.070
0.6	0.253	0.228	0.142	-0.002

Table E2.7 Biaxiality ratio according to eq.(E2.7.10) for 3-point bending, S/W=5.

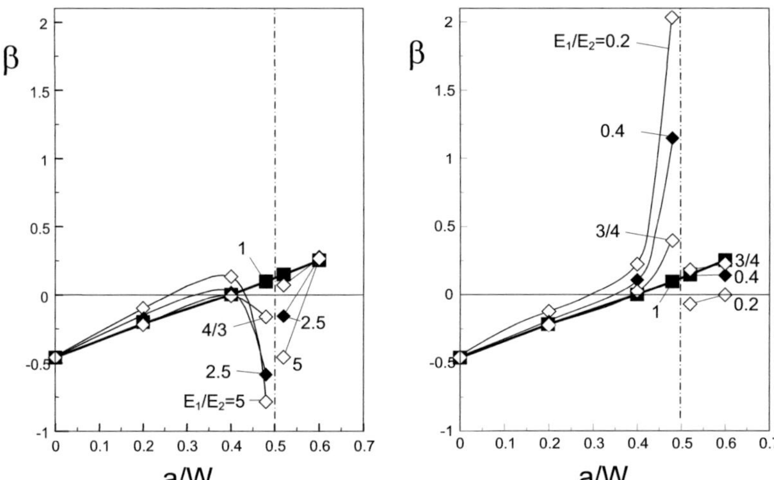

Fig. E2.40 Biaxiality ratio for 3-point bending at different ratios E_1/E_2, S/W=5.

a/W	E_1/E_2=1	1/0.75	1/0.4	1/0.2
0	-0.469	-0.469	-0.469	-0.469
0.2	-0.227	-0.209	-0.160	-0.104
0.4	0.096	0.074	0.081	0.186
0.48	0.231	-0.069	-0.553	-0.781
0.52	0.291	0.183	-0.113	-0.462
0.6	0.435	0.442	0.443	0.435

Table E2.8 Biaxiality ratio according to eq.(E2.7.10) for 4-point bending, S_2/W=2.5.

a/W	$E_1/E_2=1$	0.75	0.4	0.2
0	-0.469	-0.469	-0.469	-0.469
0.2	-0.227	-0.240	-0.252	-0.216
0.4	0.096	0.131	0.242	0.387
0.48	0.231	0.573	1.428	2.420
0.52	0.291	0.371	0.422	0.241
0.6	0.435	0.417	0.437	0.208

Table E2.9 Biaxiality ratio according to eq.(E2.7.10) for 4-point bending, $S_2/W=2.5$.

References E2:

E2.1 Rice, J.R., Elastic fracture mechanics concepts for interfacial cracks, J. Appl. Mech. **55**(1988), 98-103.

E2.2 Sih, G.C., Chen, E.P., Cracks in composite materials, Ch.3, (Mechanics of Fracture VI), ed. by G.C. Sih, Martinus Nijhoff Publishers, The Hague (1981).

E2.3 He, M.Y., Hutchinson, J.W., Kinking of a crack out of an interface, J. Appl. Mech. **56**(1989), 270-278.

E2.4 He, M.Y. Bartlett, A., Evans, A.G., Hutchinson, J.W., Kinking of a crack out of an interface: Role of in-plane stress, J. Am. Ceram. Soc. **74**(1991), 767-771.

E2.5 Cotterell, B., Rice, J.R., Slightly curved or kinked cracks, Int. J. of Fract. **16**(1980), 155-169.

E2.6 Hutchinson, J.W., Mixed mode fracture mechanics of interfaces, in: Metal/Ceramic Interfaces, Ed. by M. Rühle, A.G. Evans, M.F. Ashby, and J.P. Hirth, Pergamon Press, Elmsford, NY, 1990.

E2.7 Murakami, Y., Hanson, M.T., Hasebe, N., Itoh, Y., Kishimoto, K., Miyata, H., Miyazaki, N., Terada, H., Tohgo, K., Yuuki, R., Stress Intensity Factors Handbook, Vol. 3. Pergamon Press, (1992), Oxford, UK.

E2.8 Fett, T., Munz, D. (1997). Stress Intensity Factors and Weight Functions, Computational Mechanics Publications, Southampton, UK.

E2.9 Fett, T., Rizzi, G., T-stress solutions determined by FE computations, Report FZKA 6937, Forschungszentrum Karlsruhe, Karlsruhe (2004)

E2.10 Srawley, J.E., Wide range stress intensity factor expressions for ASTM E399 Standard Fracture Toughness Specimens, Int. J. Fract. Mech. **12**(1976), 475-476.

E2.11 He, M.Y., Turner, M.R., Evans, A.G., Analysis of the double cleavage drilled compression specimen for interface fracture energy measurements over a range of mode mixities, Acta metall. mater. **43**(1995), 3453-3458.

E2.12 Ritter, J.E., Huseinovic, A., Chakravarthy, S., Lardner, T.J., Subcritical crack growth in soda-lime glass under mixed-mode loading, J. Amer. Ceram. Soc. **83**(2000), 2109-2111.

E2.13 Lardner, T.J., Chakravarthy, S., Quinn, J.D., Ritter, J.E., Further analysis of the DCDC specimen with an offset hole, Int. J. Fract. **109**(2001), 227-237.

E2.14 Fett, T., Rizzi, G., Munz, D., T-stress solution for DCDC specimens, Engng. Fract. Mech. **72**(2005), 145-149.

E2.15 Suga, T., Bruchmechanische Charakterisierung und Bestimmung der Haftfestigkeit von Materialübergängen, Thesis 1983, Stuttgart

E2.16 Fett, T., Munz, D., Thun, A toughness test device with opposite roller loading, Engng. Fract. Mech. **68**(2001), 29-38.

E2.17 Fett, T., T-stress and stress intensity factor solutions for 2-dimensional cracks, VDI-Verlag, 2002.

E2.18 Noda, N.A., Araki, K., Erdogan, F., Stress intensity factors in two bonded elastic layers with a single edge crack under various loading conditions, Int. J. Fract. **57**(1992), 101-126.

E2.19 Murakami, Y., et al., Stress intensity factors handbook, Pergamon Press, 1986.

E2.20 Lu, M.C., Erdogan, F., Stress intensity factors in two bonded elastic layers containing cracks perpendicular to and on the interface – II. Solution and results, Engng. Fract. Mech. **18**(1983), 507-528.

E2.21 Fujino, K., Sekine, H., Abe, H., Analysis of an edge crack in a semi-infinite composite with a long reinforced phase, Int. J. Fract. **25**(1984), 81-94.

E2.22 Rizk, A.A., Erdogan, F., Cracking of coated materials under transient thermal stresses, J. of Therm. Stresses **12**(1989), 125-168.

E2.23 Fett, T., Diegele, E., Munz, D., Rizzi, G., Weight functions for edge cracks in thin surface layers, Int. J. Fract. **81**(1996), 205-215.

E2.24 Cook, T.S., Erdogan, F., Int. J. of Engng. Sci. **10**(1972), 677-697.

E2.25 Eischen, J.W., Fracture of nonhomogeneous materials, Int. J. Fract. **34**(1987), 3-22.

E2.26 Ang, W.T., Clements, D.L., On some crack problems for inhomogeneous elastic materials, Int. J. Solids&Struct. **23**(1987), 1089-1104.

E2.27 Erdogan, F., Fracture problems in FGM and application to graded interfaces and coatings, Proc. Fist Intern. Symp. FGM, Sendai 1990, 307-312.

E2.28 Erdogan, F., Wu, B.H., Analysis of FGM specimens for fracture toughness testing, in: Ceramic Transactions, Vol. 34: Functionally Gradient Materials (ed. by J.B. Holt et al.), 47-54.

E2.29 Jin, Z.-H., Noda, N., Transient thermal stress intensity factors for a crack in a plate of a functionally gradient material, Int. J. Solids&Struct. **31**(1994), 203-218.

E2.30 Bleek, O., Munz, D., Schaller, W., Yang, Y.Y., Effect of a graded interlayer on the stress intensity factor of cracks in a joint under thermal loading, Engng. Fract. Mech. **60**(1998), 615-623.

E2.31 Leevers, P.S., Radon, J.C., Inherent stress biaxiality in various fracture specimen geometries, Int. J. Fract. **19**(1982), 311-325.

E3

Straight cracks ahead of slender notches

E3.1 Geometry and stress intensity factor

A specimen containing a slender edge notch of depth a_0 with the notch root radius R is considered (Fig. E3.1a). Such notches are often used in material testing of ceramics to simulate starter cracks. A small crack of length ℓ is assumed to emanate directly at the notch root. Stress intensity factors and T-stresses for this configuration were given in Section C19 of [E3.1]. In this section now the compliance and the weight function are considered.

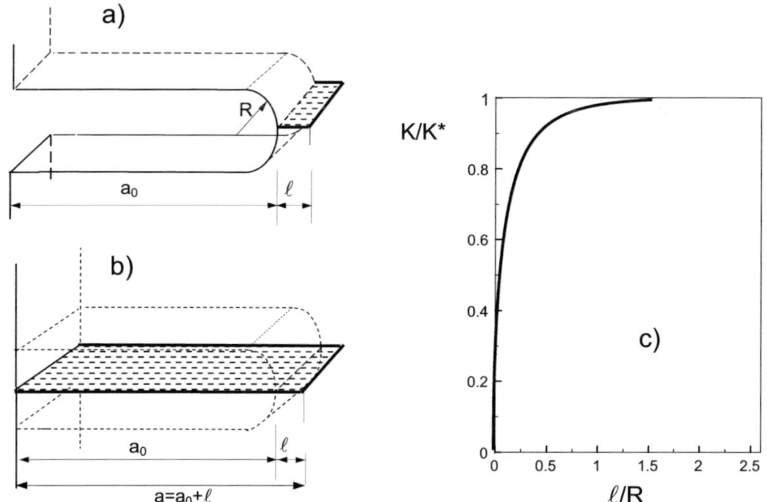

Fig. E3.1 a) Crack of length ℓ ahead of a slender notch with notch root radius R, b) same crack/notch configuration replaced by an auxiliary crack of total length $a=a_0+\ell$, c) true stress intensity factor K normalized on the formally computed value K^*.

In the "long-crack approach" the configuration of notch with length a_0 and crack of length ℓ (Fig. E3.1b) is assumed to be equivalent to a crack of total length $a = a_0+\ell$. The related stress intensity factor is *formally* computed as

$$K^* = \sigma\sqrt{\pi(a_0 + \ell)}\,F(a/W),\tag{E3.1.1}$$

where F is the geometric function for an edge crack of depth a in a specimen of width W. This geometric function is known for many test specimens and available from fracture mechanics handbooks. The stress intensity factor K^* is sufficiently correct only for cracks with $\ell \gg R$.

In the first crack extension phase, however, the crack length ℓ is comparable to R. The quantity K^* then strongly deviates from the correct stress intensity factor K. If the notch root radius is small compared to the notch depth a_0 and other specimen dimensions, the true stress intensity factor K is given by [E3.2]

$$K/K^* \cong \tanh[A\sqrt{\ell/R}] \tag{E3.1.2}$$

(A=2.243). This relation is shown in Fig. E3.1c by the curve. From this plot it is clearly visible that the true stress intensity factor K is significantly lower than the formally computed K^* for the very first crack extension. On the other hand, it can be concluded from Fig. E3.1c that notch effects are negligible if $\ell > 1.5R$.

E3.2 Compliance

In bending, the increase of the formally computed compliance C^* due to the existence of a small crack of length ℓ is given by

$$\Delta C^* = \frac{9}{2}\frac{L^2\pi}{BW^4E'}\int_{a_0}^{a_0+\ell}F^2\,a'\mathrm{d}(a') \tag{E3.2.1}$$

with the supporting span L in 3-point bending or $L=S_1-S_2$ in 4-point bending (S_1=supporting and S_2= loading roller span). The "long-crack solution" ΔC^* becomes incorrect if the condition $\ell \gg R$ is not fulfilled.

The compliance of a crack in front of a slender notch is given as [E3.3]

$$\Delta C = \frac{9}{2}\frac{L^2\pi}{BW^4E'}\int_{a_0}^{a_0+\ell}F^2\,(a_0+\ell)'\tanh^2[A\sqrt{\ell'/R}]\mathrm{d}(\ell') \tag{E3.2.2}$$

with an approximate analytical solution:

$$\Delta C \cong \left(1+\frac{2}{A^2}\frac{R}{\ell}\ln(\cosh[A\sqrt{\ell/R}])-\frac{2}{A}\sqrt{R/\ell}\,\tanh[A\sqrt{\ell/R}]\right)\Delta C^* \tag{E3.2.3}$$

For the computation of ℓ from ΔC, eq. (E3.2.2) has to be solved with respect to the upper integration limit. For this purpose the program module "FindRoot" of Mathematica [E3.4] can be used with the upper integration limit ℓ in (E3.2.2) taken as the unknown quantity or by solving (E3.2.3) numerically. The compliance increment ΔC from (E3.2.2) normalised on the formally computed value ΔC^* from (E3.2.1) is plotted in Fig. E3.2. Although the two individual compliances ΔC and ΔC^* depend strongly on the notch length a_0, the numerical evaluation shows that the ratio of

$\Delta C/\Delta C^*$ is nearly unaffected by a_0. From Fig. E3.2 it can be concluded that for crack extensions in the order of a few notch root radii ($\ell>4R$), the formally computed compliance will represent the notch/crack-configuration with sufficient accuracy.

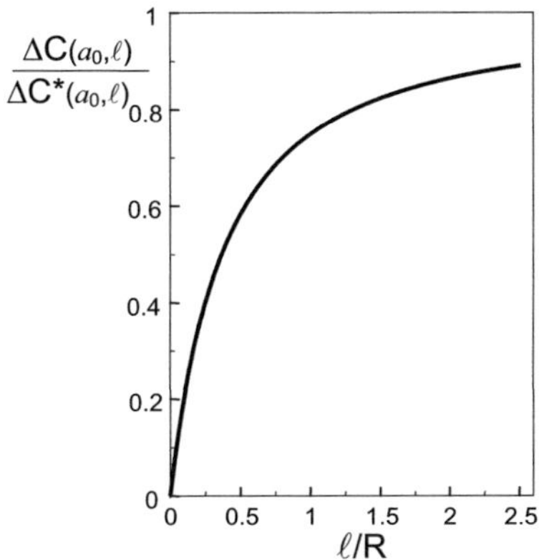

Fig. E3.2 True compliance increase ΔC normalised on the „long-crack solution" ΔC^* for the same crack increment.

E3.3 Weight function

For computations of bridging stress intensity factors for ceramics with crack-surface interactions it is necessary to know the weight function for the crack ahead of the notch root (geometric data in Fig. E3.3a). For this purpose a simple interpolation relation was given in [E3.2] as

$$h_{notch} = \lambda h^{(2)} + (1-\lambda)h^{(1)} \quad , \quad \lambda = \left(\frac{R}{R+\ell}\right)^{7/2} \tag{E3.3.1}$$

where
$$h^{(1)} = h_{edge}\left(\frac{\xi+a_0}{a},\frac{a}{W}\right), \; \ell = a - a_0 \tag{E3.3.2}$$

is the "long-crack weight function" for a crack of total length a (see Fig. E3.3b) and

$$h^{(2)} = h_{edge}\left(\frac{\xi}{\ell},\frac{\ell}{W-a_0}\right) \tag{E3.3.3}$$

is the weight function of an edge crack in a component of reduced width W-a_0 (see Fig. E3.3b). The weight function can be written for the special case of a small crack extension compared to the initial crack length, $\ell \ll a_0$

$$h_{notch} = \sqrt{\frac{2}{\pi \ell}} \left[\frac{1}{\sqrt{(\ell-\xi)/\ell}} + 0.568\,\lambda \left(\sqrt{\frac{\ell-\xi}{\ell}} + \tfrac{1}{2}\left(\frac{\ell-\xi}{\ell}\right)^{3/2} \right) \right] \qquad \text{(E3.3.4a)}$$

with
$$\lambda = \left(\frac{R}{R+\ell}\right)^{7/2}, \quad \ell = a - a_0 \qquad \text{(E3.3.4b)}$$

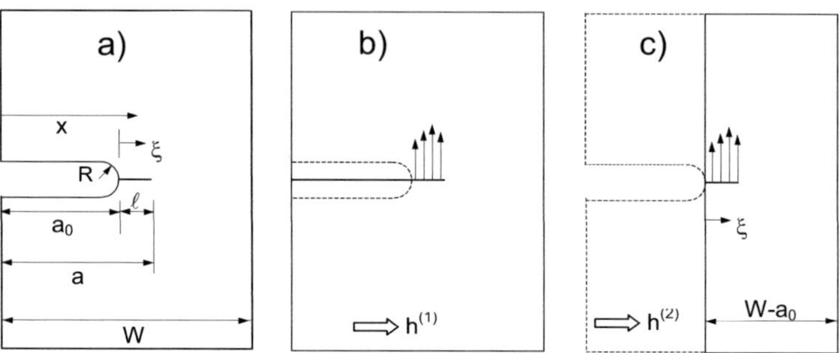

Fig. E3.3 a) Geometric data necessary for computing the weight function for a crack emanating from a notch root, b) "long-crack approach" $h^{(1)}$ represented by an edge crack of total length $a=a_0+\ell$, c) limit case $h^{(2)}$: an edge crack of depth ℓ in a plate or bar of reduced width $W-a_0$.

E3.4 Visibility of cracks emanating from narrow notches

Crack length measurements on fracture mechanics test specimens containing a narrow notch are often carried out by observation of the crack tip on the side surface using an optical microscope. At the beginning of a test, such notches show not any crack at the side surface although the increasing compliance clearly indicates crack extension. A reason for this may be given here.

A slender notch of initial depth a_0 and root radius R in a mechanically loaded specimen is considered (Fig. E3.4a). The normal stresses σ_y ahead of the notch root (Fig. E3.4b) can be computed according to Creager and Paris [E3.5] by

$$\sigma_y = \frac{2K(a_0)}{\sqrt{\pi(R+2\xi)}} \frac{R+\xi}{R+2\xi} \qquad \text{(E3.4.1)}$$

where the quantity $K(a_0)$ is the stress intensity factor of a crack having the same length a_0 (Fig. E3.4c) as the notch. It holds

$$K(a_0) = \sigma^* F(a_0)\sqrt{\pi a_0} \qquad \text{(E3.4.2)}$$

with the characteristic stress σ^* (e.g. remote tensile stress, outer fibre bending stress) and the geometric function F. The solid part of the curve in Fig. E3.4b represents the region ($0 \leq \xi \leq R/2$) in which higher-order terms in the stress approximation by Creager and Paris are negligible and (E3.4.1) sufficiently describes the stress field.

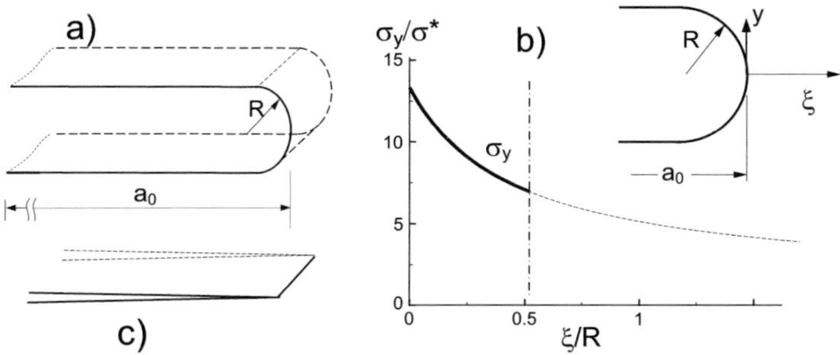

Fig. E3.4 a) A through-the-thickness slender notch, b) normal stress σ_y ahead of the notch under bending load computed according to Creager and Paris [E3.5] for $a_0/W = 0.5$ and $R/W = 0.025$; W=width of the bending bar, c) crack of size a_0 under the same load.

In order to determine the location at the notch front at which a crack will be initiated and propagate, we first have to look for the maximum of $K(a_0)$ along the crack front of the 3-dimensional crack problem of Fig. E3.4c.

It is well known in fracture mechanics that the stress intensity factor K and the energy release rate $G \propto K^2$ vary along a straight crack front. This fact is illustrated in Fig. E3.5a, where the local energy release rates G_{3D} are plotted normalised on the G-values obtained by 2D modelling assuming plane stress or plane strain conditions.

The squares show results of Dimitrov et al. [E3.6] obtained for a straight crack in a 3-point bending bar. The circles are results for a "double cleavage drilled compression" (DCDC) test specimen [E3.7]. In both cases the energy release rates show a maximum in the specimen centre and significantly reduced values in the surface region.

Directly at the free side surface, $z/B \rightarrow \pm 1/2$, the description of the singular stress field by a stress intensity factor is no longer possible. In this case, the stresses are given by the more general relation

$$\sigma_y \propto r^{\lambda-1} \qquad (E3.4.3)$$

with $\lambda \cong 0.54$ for a crack terminating angle of $\gamma=0$ (straight crack) [E3.8]. The singularity exponent λ depends on the crack terminating angle γ and to a very slight extend on Poisson's ratio ν. Equation (E3.4.3) yields a weak singularity for the stresses $\sigma \propto r^{-0.46}$ with the consequence of a disappearing energy release rate (for details see e.g. [E3.6]). In Fig. E3.5, this result is symbolised by the arrows (note that the finite G-values at the surface are a consequence of the finite FE-mesh). A finite energy release rate, necessary for stable crack growth, is ensured only if $\gamma \cong 10°$ where

$\lambda = \frac{1}{2}$ is fulfilled (see eq.(F2.2.5)) and, consequently, a stress intensity factor exists. With other words: A crack cannot grow stably at a free surface if $\gamma \neq 10°$.

Since the stress intensity factor $K(a_0)$ describing the notch stresses is proportional to the stress intensity factor of a sharp crack of length a_0, it also holds for the notch that first crack propagation must occur near the specimen center where the highest driving forces (G, K) are present (Fig. E3.5a). In a later crack extension phase the crack reaches the surface and can grow also in this region. Only now the crack becomes visible at the side surface and can be observed with the microscope.

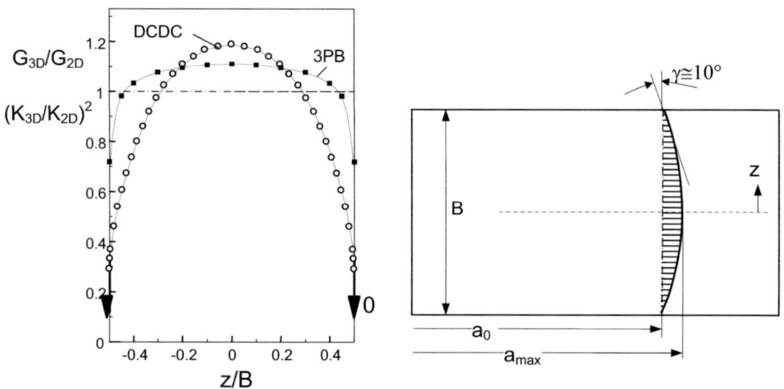

Fig. E3.5 Energy release rate distribution along the front of straight-through specimen cracks (squares: 3-point bending test [E3.6], circles: DCDC test specimen [3.7]), both results obtained from FE modelling, b) first crack development in the centre region of the notch; when an angle of $\gamma \cong 10°$ is reached, crack extension is visible also at the side surface.

References E3

E3.1 Fett, T., Stress intensity factors, T-stresses, Weight functions, IKM 50, Universitätsverlag Karls-ruhe, 2008.

E3.2 Fett, T., Munz, D., Stress Intensity Factors and Weight Functions, Computational Mechanics Publications, (1997) Southampton, UK.

E3.3 Fett, T., Notch effects in determination of fracture toughness and compliance, Int. J. Fract. **72**(1995), R27-R30.

E3.4 Mathematica, Wolfram Research Inc., USA (2003).

E3.5 Creager, M., Paris, P.C., Elastic field equations for blunt cracks with reference to stress corrosion cracking, Int. J. Fract. **3**(1967), 247-252.

E3.6 Dimitrov, A., Buchholz, F.-G., Schnack, E., "3D-corner effects in crack propagation." In H.A. Mang, F.G. Rammerstorfer, and J. Eberhardsteiner, eds, *On-line Proc. 5th World Congress in Comp. Mech. (WCCMV)*, Vienna, Austria, July 7–12(2002); http://wccm.tuwien.ac.at.

E3.7 Fett, T., Rizzi, G., Guin, J.P., López-Cepero, J.M., Wiederhorn, S.M., A fracture mechanics analysis of the DCDC test specimen, Engng. Fract. Mech. **76**(2009), 921-934.

E3.8 Benthem, J.P., "State of stress at the vertex of a quarter-infinite crack in a half-space," Int. J. Solids and Struct. **13**(1977), 479.

E4

Trapezoidal edge-cracked bar

E4.1 Test specimen

The trapezoidal bar with an edge crack has been addressed in Section D3 of [E4.1]. This specimen type is of special interest for applications in the field of ceramics. Extremely sharp starter notches may be introduced for instance by using a focussed ion beam (FIB) device [E4.2]. The notch root radius in such cases (some 10 nm) is clearly smaller than the mean grain size of most ceramics. Consequently, the starter notch acts like a sharp crack.

In [E4.1] stress intensity factor solutions obtained by FE-computations were reported in the form of diagrams. In this Section tables for interpolations and approximate relations are given for K and in addition the compliance is reported.

Figure E4.1 shows a test specimen. The thickness at the tensile side of a trapezoidal bar, b_0, is in the order of 0.5mm, the thickness B at the compression surface is identical with the standard thickness of rectangular bending bars stand on end ($B \approx 3$mm). Also the width W is a standard dimension of bending bars ($W \approx 4$mm). In order to ensure stability during load application, the outer specimen parts may remain rectangular.

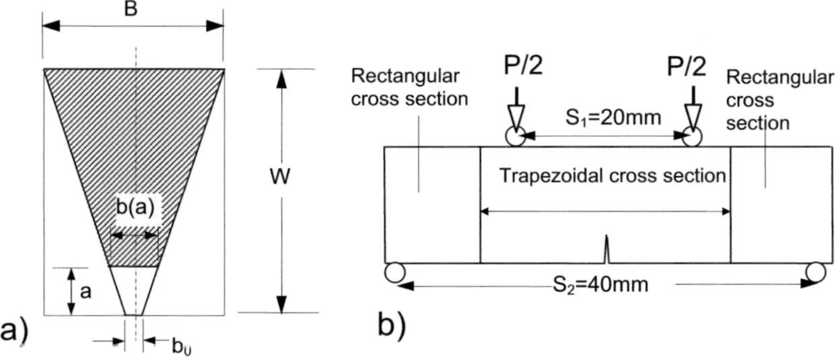

Fig. E4.1. A 4-point bending test specimen with a pre-notched trapezoidal bar, a) cross section, b) loading situation.

The bending stress defined by the outer fibre tensile stress for a trapezoidal test specimen is in 4-point bending

$$\sigma_{bend} = 3\frac{P(S_2 - S_1)}{W^2}\,\kappa \tag{E4.1.1}$$

$$\kappa = \frac{2B + b_0}{B^2 + 4Bb_0 + b_0^2} \tag{E4.1.2}$$

The distance h of the neutral axis from the tensile surface is

$$h = \frac{W}{3}\frac{2B + b_0}{B + b_0} \tag{E4.1.3}$$

E4.2 Stress intensity factors

The stress intensity factors for several relative crack depths $\alpha=a/W$ and thickness ratios b/B were computed in [E4.2] by a 3-dimensional finite element study.

The stress intensity factor for a fixed $B/W=3/4$ and various a/W and b_0/B is represented in Fig. E4.2 by the geometric function F defined by

$$K = \sigma_{bend}\sqrt{\pi a}\,F \tag{E4.2.1}$$

Table E4.1 compiles the F-results which may be interpolated with respect to a/W and b_0/B.

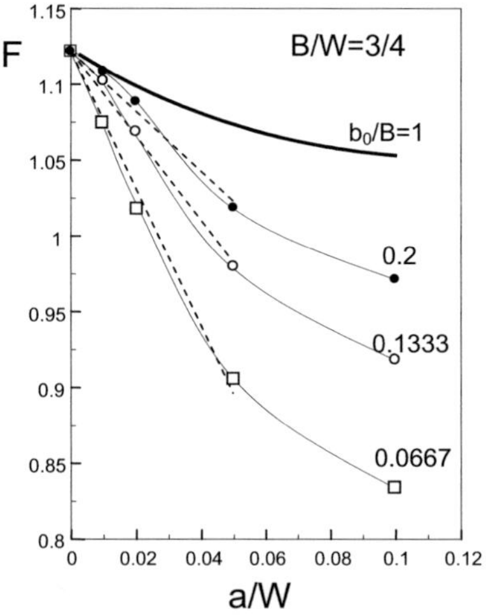

Fig. E4.2 Geometric function F for stress intensity factors; linear approximation of data for $a/W \leq$ 0.05 (dashed lines), $\nu = 0.25$.

For limited ranges of $a/W \leq 0.05$ and $0.067 \leq b_0/B \leq 1$, the FE results can be approximated by

$$F \cong F_0\left(1 - \lambda\frac{a}{W}\right), \quad F_0 = 1.1215, \quad \lambda = 0.838 + 0.216\frac{B}{b_0} \qquad (E4.2.2)$$

with the theoretical limit F_0 representing the edge-cracked half-space. The deviations are less than 2%. The approximation is represented by the dashed straight lines in Fig. E4.2.

b_0/B	$a/W=0$	0.01	0.02	0.05	0.1
1	1.12155	1.112	1.098	1.0721	1.0519
1/5	1.12155	1.103	1.084	1.0185	0.9711
2/15	1.12155	1.096	1.063	0.9796	0.9181
1/15	1.12155	1.073	1.017	0.9043	0.8340

Table E4.1 Geometric function F according to eq.(E4.2.2).

E4.3 Compliance

The compliance caused by the crack exclusively is given by

$$C = \frac{2}{E'}\int_0^a \frac{1}{P^2}K^2 b(a')\,da' \qquad (E4.3.1)$$

with the effective modulus E'

$$E' = \begin{cases} E & \text{for plane stress} \\ E/(1-v^2) & \text{for plane strain} \end{cases} \qquad (E4.3.2)$$

(E= Young's modulus, v=Poisson's ratio).
The actual crack front length, $b(a)$, is given by

$$b(a) = b_0 + \frac{B-b_0}{W}a \qquad (E4.3.3)$$

The numerical compliance results are compiled in Table E4.2 in the normalized form of

$$C = \frac{9\pi(S_2 - S_1)^2}{E'W^2}\kappa^2 F_0^2 b_0\, \hat{C}(a/W, b_0/B) \qquad (E4.3.4)$$

with

$$\hat{C} = \frac{2}{F_0^2} \int\limits_0^\alpha F^2(\alpha') \frac{b(\alpha')}{b_0} \alpha' \, d\alpha' \qquad\qquad (E4.3.5)$$

b_0/B	$a/W=0$	0.01	0.02	0.05	0.08	0.1
1	1	0.986	0.972	0.937	0.913	0.902
1/5	1	1.005	1.007	1.001	0.999	1.009
2/15	1	1.012	1.016	1.013	1.021	1.041
1/15	1	1.032	1.047	1.072	1.120	1.171

Table E4.2 Normalized compliance $\hat{C}$ according to eq.(E4.3.5).

References E4

[E4.1] Fett, T., Stress intensity factors, T-stresses, Weight functions, IKM 50, Universitätsverlag Karlsruhe, 2008.
[E4.2] Fett, T., Creek, D., Wagner, S., Rizzi, G., Volkert C., Fracture toughness test with a sharp notch introduced by focussed ion beam, Int. J. Fract. **153**(2008), 85-92.

E5

DCDC test specimen

E5.1 Side-surface displacements

In recent studies to address the question of plastic deformation zones at crack tips in silicate glasses, cracks emerging from the side surfaces of fracture mechanics test specimens were examined using atomic force microscopy [E5.1, E5.2]. Specimen surfaces surrounding the crack were found to be depressed. In order to distinguish plastic effects from the normal linear-elastic depression field near crack tips, a 3-dimensional finite element analysis on DCDC-specimens (Fig. E5.1) was carried out in [E5.3].

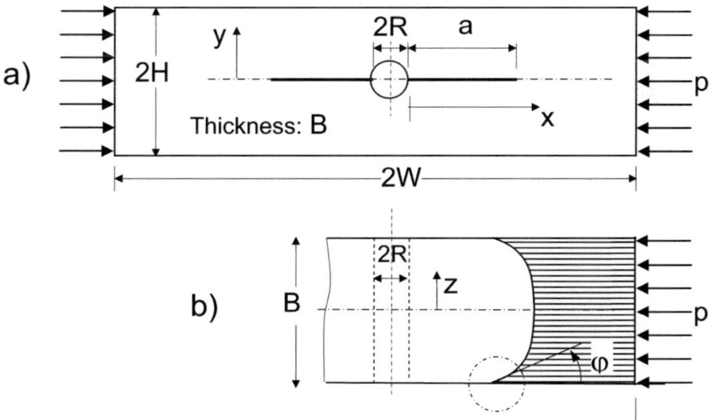

Fig. E5.1 DCDC specimen, loaded by compressive stresses p at the ends, a) side view, b) view on the crack plane.

For different crack terminating angles the displacements along the polar angles $\theta = 0°$ (along $y = 0$, $x \geq a$), 90° (along axis $x = a$) and 180° (along $y = 0$, $x < a$) were fitted over relative crack-tip distances of $0 < |(x - a)/H| \leq 0.002$ and $0 < |y/H| \leq 0.002$ as functions of the distance r from the crack tip by

$$\frac{\Delta u_z E}{K_I \sqrt{H}} = \sum_{(n)} A_n \left(\frac{r}{H}\right)^{q_n} \tag{E5.1.1}$$

$$r = \sqrt{(x - a)^2 + y^2}$$

with the coefficients A_n compiled in Table E5.1 and the actual stress intensity factor K_I according to eq.(E5.1.1). By an interpolation of the displacements with respect to the tabulated polar angles, θ, the displacement for any other polar angle can be estimated. The limits for $(x-a)/H$ and y/H given before are necessary in order to ensure the possibility of scaling the displacements by K. In Section E5.2 it will be shown that no influence of the T-stress term can occur. The following results were determined for the special case of a DCDC specimen with $a/R=4$.

Since the depression profile must be symmetric with respect to $\theta = 0$, the interpolation function must exhibit even exponents in θ^{2n}, exclusively. From the 3 curves we can derive the 3-terms polyomial,

$$\frac{\Delta u_z E}{K_1 \sqrt{H}} = A + B\left(\frac{\theta}{\pi}\right)^2 + C\left(\frac{\theta}{\pi}\right)^4 , \tag{E5.1.2}$$

(for θ in radians). The coefficient A is simply given as the Δu_z curve for $\theta = 0$

$$A = \left.\frac{\Delta u_z E}{K_1 \sqrt{H}}\right|_{\theta=0} . \tag{E5.1.3}$$

The other coefficients are

$$B = -5A + \frac{16}{3}\left.\frac{\Delta u_z E}{K_1 \sqrt{H}}\right|_{\theta=90°} - \frac{1}{3}\left.\frac{\Delta u_z E}{K_1 \sqrt{H}}\right|_{\theta=180°} , \tag{E5.1.4}$$

and

$$C = 4A - \frac{16}{3}\left.\frac{\Delta u_z E}{K_1 \sqrt{H}}\right|_{\theta=90°} + \frac{4}{3}\left.\frac{\Delta u_z E}{K_1 \sqrt{H}}\right|_{\theta=180°} . \tag{E5.1.5}$$

n	θ		$\varphi=45°$		$\varphi=60°$		$\varphi=80°$		$\varphi=90°$
		q_n	A_n	q_n	A_n	q_n	A_n	q_n	A_n
1	0°	0.36	0.1636	0.42	0.1612	1	-0.477	0.54	-0.112
2		1	-0.518	1	1.980	2	174.6	1	3.541
3		3/2	-15.67	3/2	-34.46			3/2	-27.18
1	90°	0.36	0.4134	0.42	0.459	1	3.132	0.54	0.00899
2		1	4.21	1	0.505	2	-365.1	1	3.747
3					0			3/2	-38.73
1	180°	1	13.53	1	10.68	1	2.812	1	14.17
2		3/2	-58.9	3/2	164.5			3/2	-333.03
3				2	-2747			2	2441.

Table E5.1 Exponents and coefficients of the displacement representation according to eq.(E5.1.1) for $r/H \leq 0.002$.

For the case of $\varphi = 90°$, additional displacements for the angles $\theta = 45°$ and $135°$ obtained by FE are plotted in Fig. E5.2a together with the prediction on the basis of the angles of $\theta = 0°$, $90°$ and $180°$. Figure E5.2b shows the additional FE-points for the two selected crack-tip distances as solid circles. At $45°$ the agreement is excellent, at $135°$ only small deviations are visible.

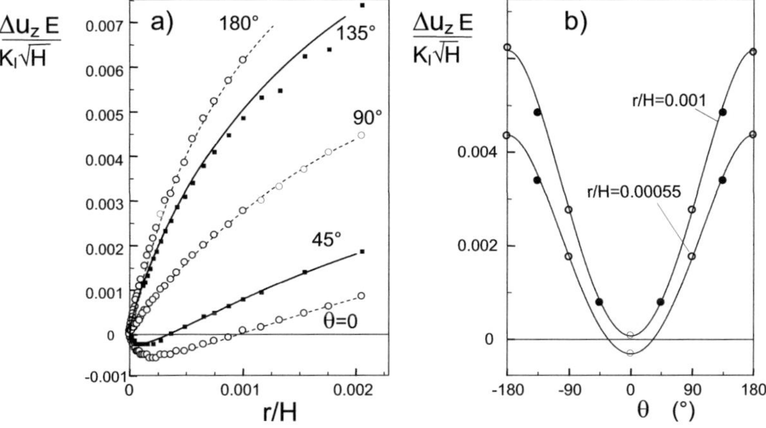

Fig. E5.2 Displacement profiles for $\varphi=90°$, a) predicted profiles for $\theta=45°$ and $135°$ (solid curves) compared with FE results (squares), b) predictions as curves, FE-results as solid circles.

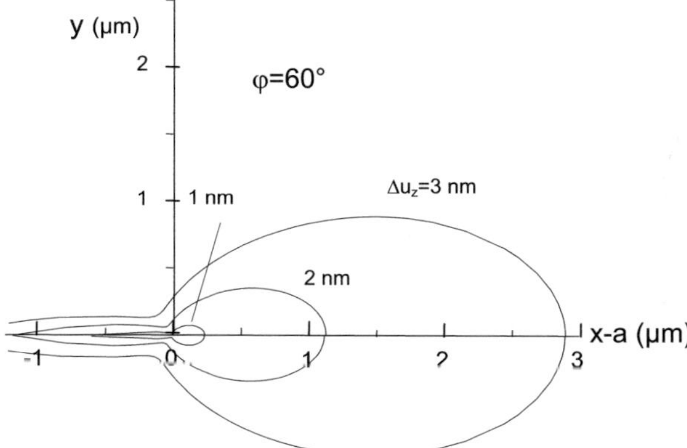

Fig. E5.3 Displacement contours for $p = 70$ MPa ($K=0.42$ MPa$\sqrt{m}$), $E = 70$ GPa, $W = 20$mm. Since the tip of the crack is defined as $\Delta u_z = 0$, displacements relative to the crack tip increase with distance from the crack tip. The gradient, however, decreases.

As an application of eqs. (E5.1.2)-(E5.1.5), the depression contours are plotted in Fig. E5.3 for $\varphi = 60°$ and the data for a glass specimen of $E = 70$ GPa, $K = 0.42$ MPa√m, $W = 20$ mm, and $B/2 = H = 0.1W$. A strong concentration of contour lines is visible near the origin $x\text{-}a = y = 0$. But this holds also for section lines in the crack wake $x < a$ parallel to the y-axis.

E5.2 Effect of T-stress on side-surface displacements

In Section E5.1 the z-displacements Δu_z were scaled on the stress intensity factor in order to allow the results for $a/R=4$ to be transferred to other crack lengths. The reason for this may be shown by considering the first higher-order stress term T.

From Hooke's law it results for the contribution of T to the strains in depth direction

$$\varepsilon_{z,T} = -\frac{\nu}{E}T \qquad (\text{E5.2.1})$$

Since $T<0$, an expansion of the DCDC specimen must occur. The z-displacements result from (E5.2.1) by integration over the half thickness as

$$u_{z,T} = -\frac{\nu}{E}\int_0^{B/2} T(z)\,dz \qquad (\text{E5.2.2})$$

where also the slight z-variation of T (see Fig. D4.2b in [E5.4]) is taken into account. In this context it has to be emphasized that the T-stress for a given pressure p at the specimen ends is a function of the crack length exclusively (but not on the coordinates x and y). Consequently, the interesting quantity $\Delta u_{z,T}$ is trivially

$$\Delta u_{z,T}(x,y) = u_{z,T}(T,x,y) - u_{z,T}(T,0,0) = 0 \qquad (\text{E5.2.3})$$

i.e. the displacements Δu_z can be scaled by the stress intensity factor K exclusively.

Having this in mind, we can apply the displacement field of Section E5.1 as a good approximation to any crack and test specimen.

E5.3 Influence of residual stresses at the side-surface

Surfaces of ceramics and glass are often affected by residual stresses. Such stresses may be caused by surface treatment as polishing or by reaction of water with the glass (e.g. ion exchange) at the surface of test specimens. In Section C22 of [E5.4], the influence of residual stresses on stress intensity factors was studied for the special case of volume changes along the crack surfaces. Such water-affected layers will occur also on the side surfaces influencing the local stress intensity factors.

In a 3-dimensiona finite element analysis, the local stress intensity factor due to a residual stress layer at the side surfaces of a DCDC specimen was determined. The crack in this specimen was modelled with a straight front (straight through the thickness B, see Fig. E5.4a). The layer thickness d was chosen to be $d=10^{-3}(B/2)$. This special choice sufficiently represents the cracked half-space with a residual stress surface layer.

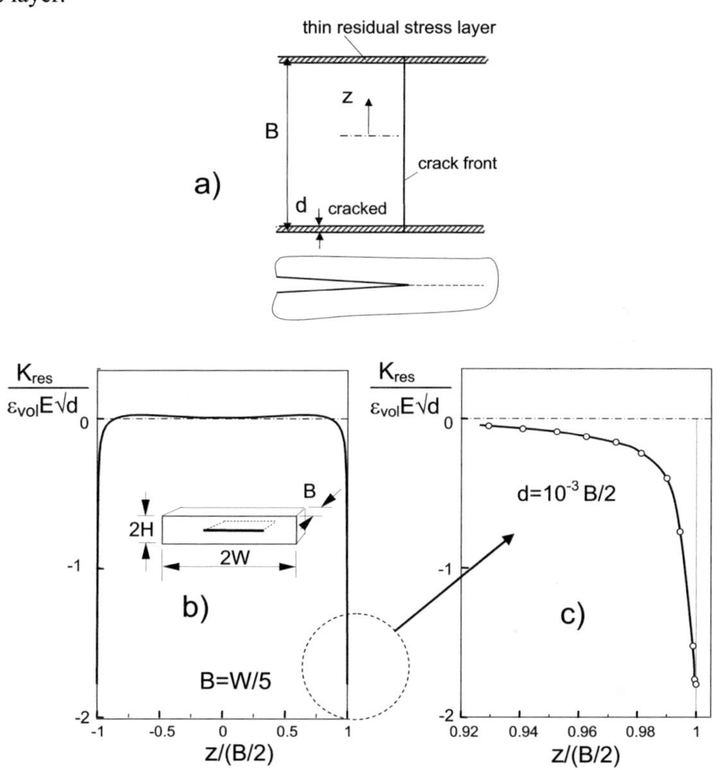

Fig. E5.4 a) Residual stress layers at the free specimen surfaces e.g. due to ion exchange, b) residual stress intensity factor obtained by 3-dimensional finite element computations, c) detail of b) near the surface.

The residual stress in the layers is

$$\sigma_{layer} = -\frac{\varepsilon_{vol} E}{3(1-\nu)}$$ (E5.3.1)

with the volumetric strain in the layer ε_{vol}, the Young's modulus E, and Poisson's ratio ν chosen to be $\nu=0.25$.

The related residual stress intensity factors K_{res} are plotted in Fig. E5.4b (and in more detail in Fig. E5.4c).

The maximum compressive residual stress intensity factor in Fig. E5.4 is about

$$K_{res,max} \approx -1.8\,\varepsilon_{vol}E\sqrt{d} \approx 5.4(1-v)\sigma_{layer}\sqrt{d} \qquad (E5.3.2)$$

The residual stress intensity factor K_{res} normalized on the maximum value $K_{res,max}$ is plotted in Fig. E5.5a as a function of the distance from the surface ζ normalized on the layer thickness d

$$K_{res} = \frac{K_{res,max}}{\exp(-0.316\zeta/d)+0.417\zeta/d} \qquad (E5.3.3)$$

For an arbitrary residual stress distribution the residual stress intensity factor can be computed by replacing the continuous stress by zones with constant stresses according to Fig. E5.5b.

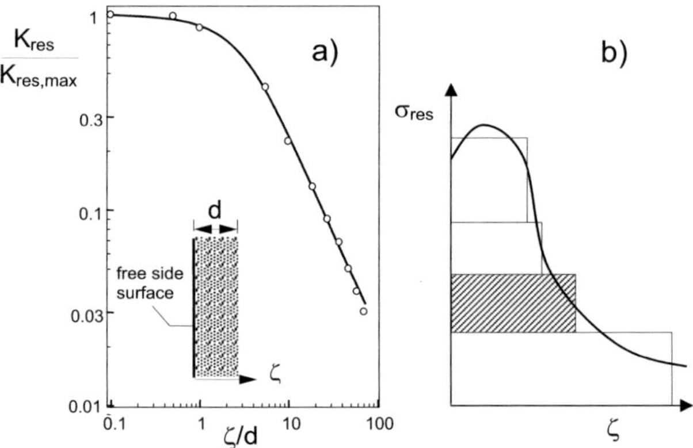

Fig. E5.5 a) Residual stress intensity factor distribution in normalized representation (ζ=distance from the surface), b) residual stress distribution replaced by zones with constant stresses.

E5.4 Stress intensity factor for a sectional straight crack front

If compressive stresses (expansive strains) occur at the side surfaces, the actual crack front in a crack growth test under superimposed external load must stay behind (Fig. E5.6). At these locations, the stress intensity factor by the external load is increased. This results in a curved crack front.

A 3-dimensional FE study was performed for computing the stress intensity factor under external load (see Section D4 of [E5.4]. Figure E5.7a shows the crack approximated by straight segments. The outer crack part intersects the free surface under an angle of φ (φ=90° corresponds to the straight crack). The next deeper part was modelled as a straight line with an intermediate angle of (φ+90°)/2. The geometric function of the local stress intensity factor through the whole specimen was plotted in Fig. D4.5b for φ=90°, 60°, and 45°.

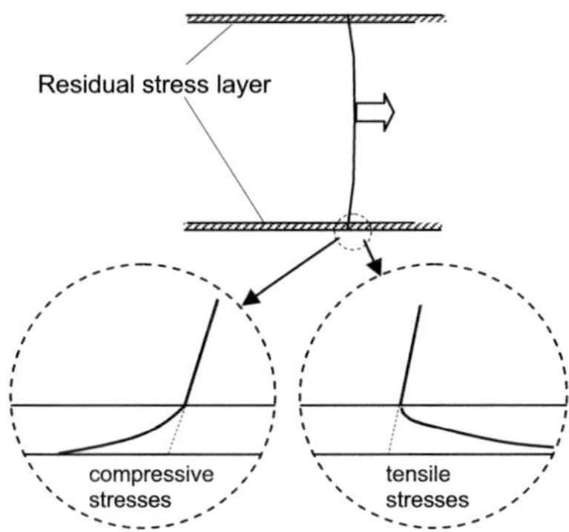

Fig. E5.6 Curved crack front in a thin surface layer caused by residual stresses.

The near-surface results of the geometric function F, defined by

$$K = \mid p \mid F\sqrt{\pi R}$$
(E5.4.1)

are plotted in Fig. E5.7b. It has to be noted that for φ≠78-82° a stress intensity factor cannot exist directly at the free surface as outlined in Sections E3.4 and F2.2. The FE-results represent an average K-value over the last few elements near the surface. Therefore, these data represent an *extrapolation* to $\zeta \rightarrow 0$.

The near-surface stress intensity factors $K_{\zeta \rightarrow 0}$ of Fig. E5.7b are plotted in Fig. E5.8 including the trivial limit cases of $K(\varphi \rightarrow 0) = \infty$ and $K(\varphi \rightarrow \pi) = 0$.

The data of Fig. E5.8a (valid for the specially chosen geometry) were fitted by

$$F_{\zeta \rightarrow 0} \cong 0.072[\tan \tfrac{1}{2}(\pi - \varphi)]^{5/3}$$
(E5.4.2)

and represented in Fig. E5.8a and 5.8b by the curves.

From dimensional reasons we can conclude that the effect of a retarded crack zone of thickness δ on the stress intensity factor and on the geometric function in eq.(E5.4.1) should be of the form

$$F_{\zeta \rightarrow 0}(\varphi) - F_{\zeta \rightarrow 0}(\tfrac{\pi}{2}) \propto \sqrt{\delta}$$
(E5.4.3)

This allows an estimation of the geometric functions for differently deep regions in which the crack deviates from a straight one.

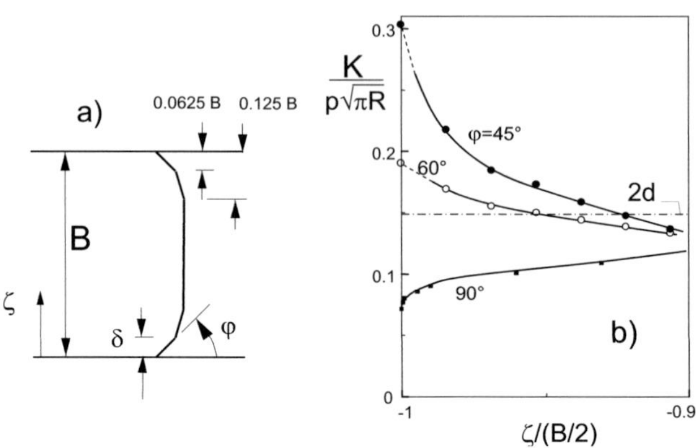

Fig. E5.7 a) Curved crack front approximated by straight segments, b) stress intensity factor near the free side surface (dash-dotted line: 2-dimensional stress intensity factor solution).

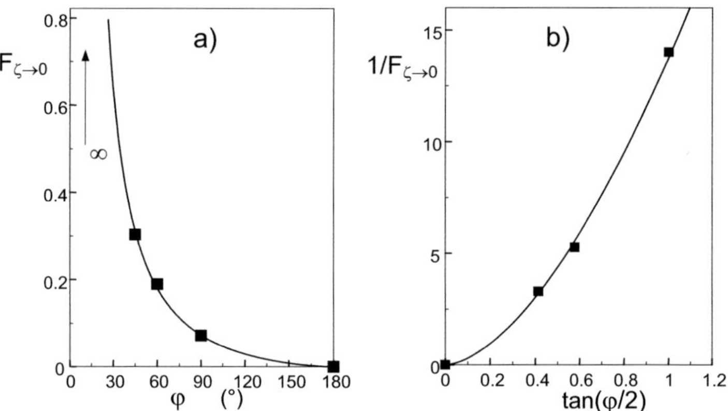

Fig. E5.8 Surface values of the geometric functions according to eq.(E5.4.1).

E5.5 Stress intensity factors for asymmetric DCDC test specimens

E5.5.1 Offset of hole and crack

In Section C15.2 of [E5.4] the asymmetric DCDC specimen with an offset b of the hole (Fig. E5.9) was addressed.

For $b/R > 0.1$ and $a/R \geq 4$, the mode-I stress intensity factors can be expressed as

$$\frac{1}{F_I} = c_0 + c_1 \frac{H}{R} + \left[c_2 \frac{H}{R} + c_3 \right] \frac{a}{R} \tag{E5.5.1a}$$

with the coefficients

$$c_0 = -0.3703 - 0.2706\,(b/R) - 0.2716\,(b/R)^2 \tag{E5.5.1b}$$

$$c_1 = 1.1163 + 0.1864\,(b/R) - 0.0140\,(b/R)^2 \tag{E5.5.1c}$$

$$c_2 = 0.2160 - 0.0326\,(b/R) + 0.0040\,(b/R)^2 \tag{E5.5.1d}$$

$$c_3 = -0.1575 + 0.0176\,(b/R) + 0.0040\,(b/R)^2 \tag{E5.5.1e}$$

where the variables b/R have to be understood as $|b/R|$.

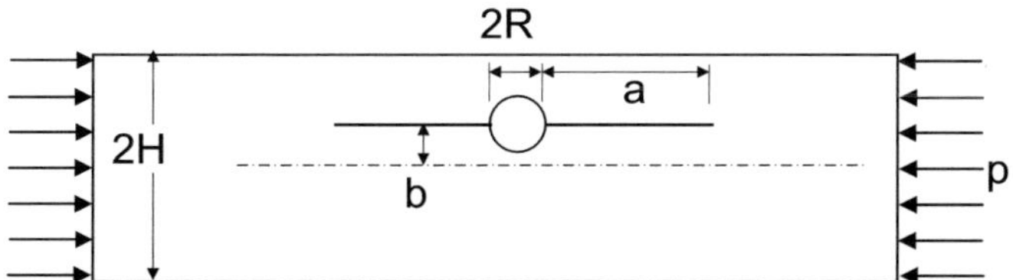

Fig. E5.9 DCDC specimen with an offset of the hole <u>and</u> the crack.

The numerical results for small offsets of $|b/R|\leq0.2$ and the commonly used geometry of $H/R=4$ can be expressed by the relation

$$F_I \cong F_{I,0}\left[1 + \left(0.122 - 0.003\frac{a}{R}\right)\left(\frac{b}{R}\right)^2\right] \tag{E5.5.2}$$

with the geometric function $F_{I,0}$ for the case of disappearing offset (see e.g. eqs. (C15.1.1) and (C15.1.2)).

E5.5.2 Offset of the hole exclusively

A second type of non-symmetry is an offset b of the hole with the crack extending in the symmetry line (Fig. E5.10). This case was studied in detail by He et al.[E5.5] and Lardner et al. [E5.6]. Their results can be expressed for $|b/R|\leq0.3$ by

$$F_I \cong F_{I,0}\left[1 + \left(3.196 - 0.0835\frac{a}{R}\right)\left(\frac{b}{R}\right)^2\right] \tag{E5.5.3}$$

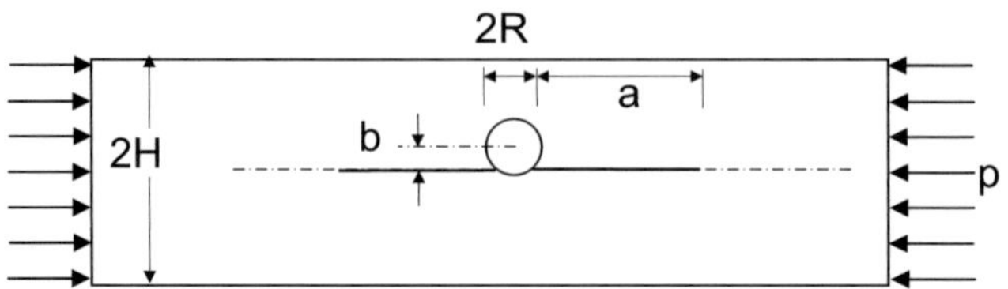

Fig. E5.10 DCDC specimen with an offset of the hole.

References E5

E5.1 Célarié, F., Prades, S., Bonamy, D., Ferrero, L., Bouchaud, E., Guillot, C., and Marlière, C., "Glass breaks like metals, but at the nanometer scale," Phys. Rev. Let., **90** [7] 075504 (2003).

E5.2 Fett, T., Rizzi, G., Creek, D., Wagner, S., Guin, J. P., Lopez-Cepero, J., Wiederhorn, S. M., Finite element analysis of a crack tip in silicate glass: No evidence for a plastic zone, PHYSICAL REVIEW B **77**, 174110 _2008

E5.3 Fett, T., Rizzi, G., Guin, J.P., López-Cepero, J.M., Wiederhorn, S.M., A fracture mechanics analysis of the double drilled compression test specimen, Engng. Fract. Mech. 76(2009), 921-934.

E5.4 Fett, T., Stress intensity factors, T-stresses, Weight functions, IKM 50, Universitätsverlag Karls-ruhe, 2008.

E5.5 He, M.Y., Turner, M.R., Evans, A.G., Analysis of the double cleavage drilled compression specimen for interface fracture energy measurements over a range of mode mixities, Acta metall. mater. **43**(1995), 3453-3458.

E5.6 Lardner, T.J., Chakravarthy, S., Quinn, J.D., Ritter, J.E., Further analysis of the DCDC specimen with an offset hole, Int. J. Fract. **109**(2001), 227-237.

E6

Mixed-Mode stress intensity factors for slant and kink cracks in finite bars

E6.1 Slant cracks under tension and bending loading

In Sections A5 and C2 of [E6.1] the slant crack in a semi-infinite body was considered. Here, the case of a finite body will be addressed.

A slant edge crack in a finite plate of width W under an angle φ is illustrated in Fig. E6.1a. The mixed-mode stress intensity factors K_I and K_{II} for such a crack may be defined via the true crack length c by

$$K_{I,II} = \sigma F_{I,II}(c)\sqrt{\pi c} \qquad (E6.1.1)$$

or by use of the crack length projection $a=c\cos(\varphi)$ via

$$K_{I,II} = \sigma F_{I,II}(a)\sqrt{\pi a} \qquad (E6.1.2)$$

with, trivially

$$F_{I,II}(c) = F_{I,II}(a)\sqrt{\cos\varphi} \qquad (E6.1.3)$$

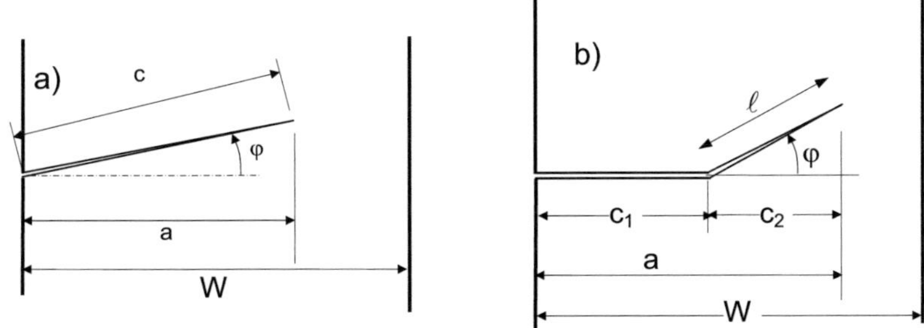

Fig. E6.1 Cracks in a bar, a) slant crack, b) kink crack.

The geometric functions $F_{I,II}$ according to [E6.2, E6.3] for $a/W{\rightarrow}0$ can be expressed by

$$F_{I,0}(c) \cong 0.5474 + 0.5738\cos(1.3225\varphi) \qquad (E6.1.4)$$

$$F_{II,0}(c) \cong 0.4741\sin(1.4675\varphi) \qquad (E6.1.5)$$

(for φ in radian). It should be noted that in this limit case the bending and tension cases are identical.

E6.1.1 Slant cracks under tensile loading
Stress intensity factors for a slant crack under bending load are given in Fig. E6.2. The circles represent data from [E6.4]. Diamond squares are results obtained by FE-analysis [E6.5]. In order to describe these data by an appropriate fit relation the kink relations of Section C3.1 may be applied.

The stress intensity factors at the tip of a kink can be expressed for $\ell=c_2/\cos(\varphi)<<a$ by

$$K_\mathrm{I}(\ell) = K_\mathrm{I}(a)\,g_{11} + b_1 T\sqrt{\ell} \qquad (E6.1.6)$$

$$K_\mathrm{II}(\ell) = K_\mathrm{I}(a)\,g_{21} + b_2 T\sqrt{\ell} \qquad (E6.1.7)$$

with the angular functions

$$g_{11} = \cos^3(\varphi/2) \qquad (E6.1.8)$$

$$g_{21} = \sin(\varphi/2)\cos^2(\varphi/2) \qquad (E6.1.9)$$

$$b_1 = \sqrt{\frac{8}{\pi}}\,\sin^2\varphi \qquad (E6.1.10)$$

$$b_2 = -\sqrt{\frac{8}{\pi}}\,\sin\varphi\cos\varphi \qquad (E6.1.11)$$

These relations are exact for kink lengths ℓ small compared to the total crack length a. Nevertheless, these solutions reflect also some main features of long kink cracks, i.e. slant cracks as limit case for $c_1\rightarrow 0$.

Fit relations for the geometric functions are

$$F_\mathrm{I}(a) = F_t[\cos^3(\tfrac{1}{2}\varphi) + 0.1504\beta\sin^2\varphi/\sqrt{\cos\varphi}] \qquad (E6.1.12)$$

$$F_\mathrm{II}(a) = F_t[\cos^2(\tfrac{1}{2}\varphi)\sin(\tfrac{1}{2}\varphi) - 0.2348\beta\sin\varphi\sqrt{\cos\varphi}] \qquad (E6.1.13)$$

with the dimensionless representation of T by the biaxiality ratio β proposed by Leevers and Radon [E6.6]

$$\beta = \frac{T\sqrt{\pi a}}{K_\mathrm{I}} \qquad (E6.1.14)$$

The function F_t denotes the tensile solution for a straight crack normal to the free side surface, given by

$$F_t = \frac{1.1215}{\lambda^{3/2}}\left[1 - 0.23566\,\lambda + \tfrac{1}{150}\lambda^2 + 3\alpha^2\lambda^7 + 0.229\exp(-7.52\tfrac{\alpha}{\lambda})\right] \quad \text{(E6.1.15)}$$

with $\alpha = a/W = c\,\cos\varphi/W$, $\lambda = 1-\alpha$.
The biaxiality ratio β for tension reads

$$\beta = \frac{-0.469 + 0.1456\,\alpha + 1.3394\,\alpha^2 + 0.4369\,\alpha^3 - 2.1025\,\alpha^4 + 1.0726\,\alpha^5}{\sqrt{1-\alpha}} \quad \text{(E6.1.16)}$$

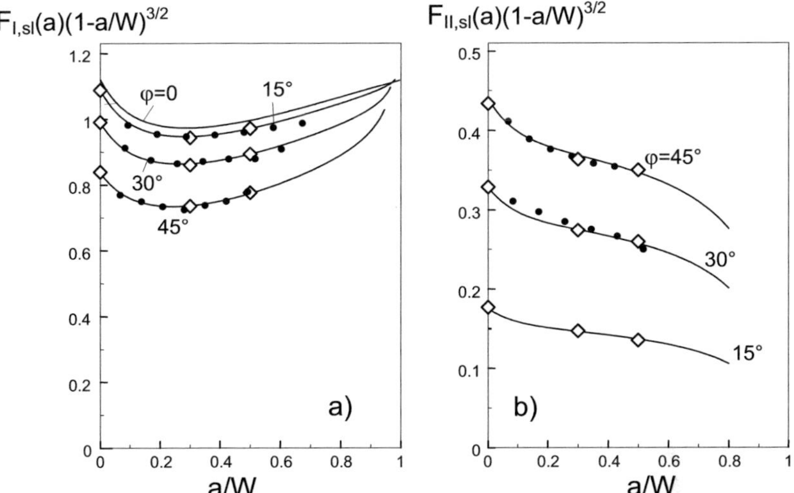

Fig. E6.2 Stress intensity factors for slant cracks under tensile loading, a) geometric function for mode I, b) geometric function for mode II; Squares: FE-results, circles: data from [E6.4], lines: fit relation according to eqs.(E6.1.12)-(E6.1.13).

E6.1.2 Slant cracks under bending loading

Stress intensity factors for a slant crack under bending load are given in Fig. E6.3. The circles represent the data from [E6.4]. Finite Element results are introduced by the diamond squares. As a fit relation it is proposed for the mode-I and mode-II geometric functions

$$F_I(a) = F_b\left[\cos^3\left(\tfrac{1}{2}\varphi\right) + 0.17845\,\beta\sin^2\varphi/\sqrt{\cos\varphi}\right] \quad \text{(E6.1.17)}$$

$$F_{\mathrm{II}}(a) = F_b[\cos^2(\tfrac{1}{2}\varphi)\sin(\tfrac{1}{2}\varphi) - 0.2182\,\beta\sin\varphi\sqrt{\cos\varphi}] \qquad \text{(E6.1.18)}$$

where F_b is the bending solution for a straight crack normal to the free side surface, given by

$$F_b = \frac{1.1215}{\lambda^{3/2}}\left[\frac{5}{8} - \frac{5}{12}\alpha + \frac{1}{8}\alpha^2 + 5\alpha^2\lambda^6 + \frac{3}{8}\exp(-6.1342\alpha/\lambda)\right] \qquad \text{(E6.1.19)}$$

again with $\alpha = a/W = c\cos\varphi/W$, $\lambda = 1-\alpha$ and the biaxiality ratio for bending

$$\beta = \frac{-0.469 + 1.2825\alpha + 0.6543\alpha^2 - 1.2415\alpha^3 + 0.07568\,\alpha^4}{\sqrt{1-\alpha}} \qquad \text{(E6.1.20)}$$

The fit functions are shown in Fig. E6.3 by the curves.

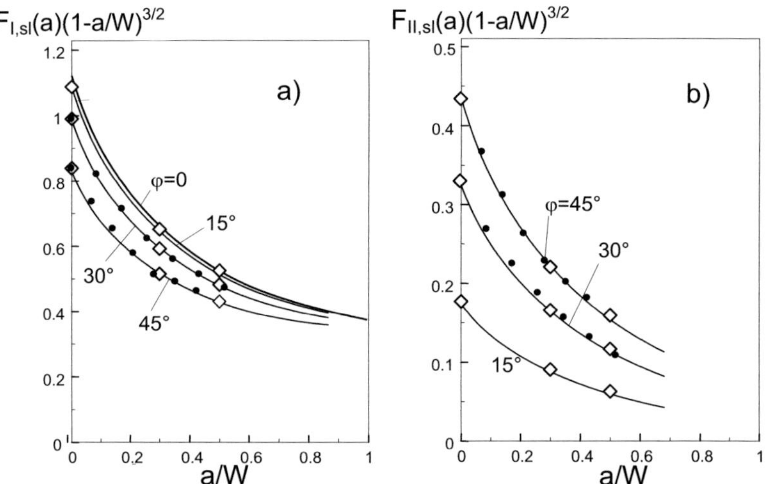

Fig. E6.3 Stress intensity factors for slant cracks under bending loading, a) geometric function for mode I, b) geometric function for mode II; Symbols: circles from [E6.4], squares: FE-computation, lines: fit relation according to eqs.(E6.1.17)-(E6.1.18).

E6.2 Kink cracks

E6.2.1 Kink cracks with infinitely small kink length

The limit case of the infinitely small kink of projected length $c_2 \to 0$ at the tip of the initial straight crack of depth c_1 (total projection of crack length: $a = c_1 + c_2$, see Fig. E6.1b) is considered first. The mixed-mode stress intensity factors are given by

$$K_{I,II,kink} = \sigma F_{I,II,kink}(a)\sqrt{a\pi} \qquad (E6.2.1)$$

with the geometric functions

$$F_{I,kink} = F_I(a)\cos^3(\varphi/2) \qquad (E6.2.2)$$

$$F_{II,kink} = F_I(a)\sin(\varphi/2)\cos^2(\varphi/2) \qquad (E6.2.3)$$

In (E6.2.2) and (E6.2.3) the geometric functions $F_I(a)$ are either the tensile or the bending solutions according to eqs.(E6.1.15) or (E6.1.19), depending on the applied load.

E6.2.2 Kink cracks with finite kink length

For $\varphi<50°$, a simple estimation of F_I and F_{II} from the limit values for $c_2/a=0$ (infinitely small kink at a 90°-crack) and $c_2/a=1$ (slant crack in bending) derived before is possible by interpolation according to

$$K_{I,II} = K_{I,II,kink} + (K_{I,II,slant} - K_{I,II,kink})\tanh[\sqrt{\mu_{I,II}\tfrac{c_2}{a}}] \qquad (E6.2.4)$$

with $\mu_I=25$ and $\mu_{II}=12$.

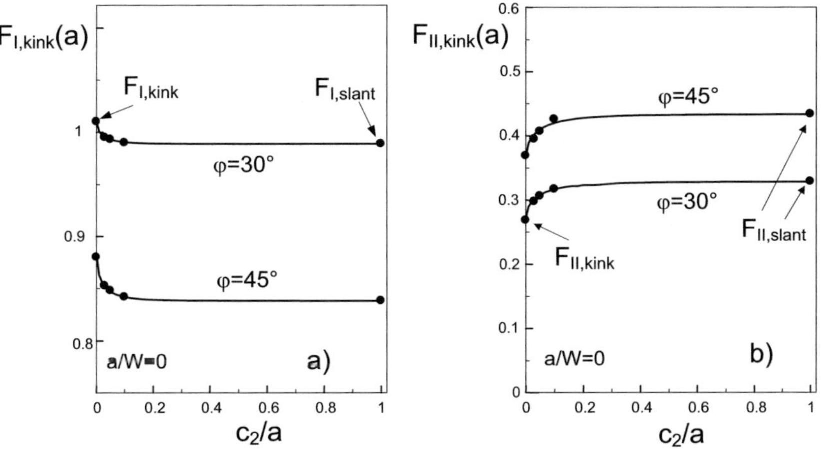

Fig. E6.4 Kink cracks for $a/W=0$ (half space); circles: data from [E6.2, E6.7], curves: interpolation according to (E6.2.4).

This interpolation relation is applied in Fig. E6.4 to literature data [E6.2, E6.7] for $\varphi=30°$ and $45°$. From the diagrams it can be concluded that already after a crack extension of c_2/a=0.15 a kink crack shows the same mixed-mode stress intensity factors as the slant crack.

Finite element results are plotted in Fig. E6.5 for tensile and bending loading. These data confirm the result of Fig. E6.4 that already for c_2/a>0.15 the kink and the slant crack exhibit identical stress intensity factors.

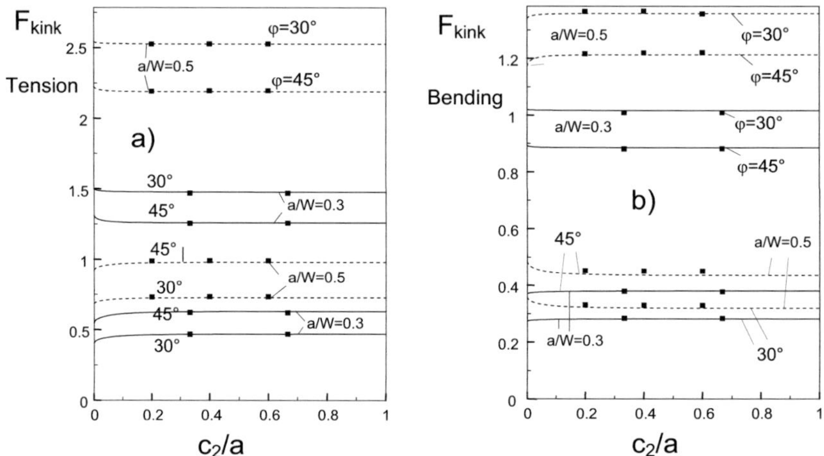

Fig. E6.5 Geometric functions for kink cracks in a finite bar under tension and bending loading; squares: FE-results, curves: interpolation according to (E6.2.4).

References E6

E6.1 Fett, T., Stress intensity factors, T-stresses, Weight functions, IKM 50, Universitätsverlag Karlsruhe, 2008.

E6.2 Noda, N.A., Oda, K., Numerical solutions of the singular integral equations in the crack analysis using the body force method, Int. J. Fract. **58**(1992), 285-304.

E6.3 Fett, T., Rizzi, G., T-stress solutions determined by finite element computations, Report FZKA 6937, Forschungszentrum Karlsruhe, 2004, Karlsruhe.

E6.4 Fett, T., Munz, D., Stress intensity factors and weight functions, Computational Mechanics Publications, Southampton, 1997.

E6.5 Bechtle, S., Fett, T., Rizzi, G., Habelitz, S., Schneider, G.A., Mixed-mode stress intensity factors for kink cracks with finite length loaded in tension and bending, to be published.

E6.6 Leevers, P.S., Radon, J.C., Inherent stress biaxiality in various fracture specimen geometries, Int. J. Fract. **19**(1982), 311-325.

E6.7 Fett, T., Rizzi, G., Weight functions for stress intensity factors and T-stress for oblique cracks in a half-space, Int. J. Fract. **132**(2005), L9-L16.

E7

Special weight function applications

E7.1 The inverse weight function problem: Stresses from K-values

The weight function considerations in [E7.1] dealt with the computation of the stress intensity factor for known stress distributions. In special applications the stresses are a priori unknown and the integral value (the stress intensity factor) is known from measurements. The problem then consists in the computation of the stress distributions. This problem has been solved in [E7.2] for fixed stresses. A more complicated problem is the determination of stresses which depend on the actual crack opening displacement as for instance occurring in the field of ceramic materials.

The increasing crack resistance of ceramic materials is of high interest for technical applications. This effect ("R-curve behaviour") is commonly described by a relation $K_R = f(\Delta a)$ in which K_R is the stress intensity factor necessary for crack propagation by an amount of Δa (Fig. E7.1a). For a survey see e.g. the article by Munz [E7.3]. Reasons for the R-curves of coarse-grained ceramics are so-called bridging stresses which act against the externally applied load (Fig. E7.1b).

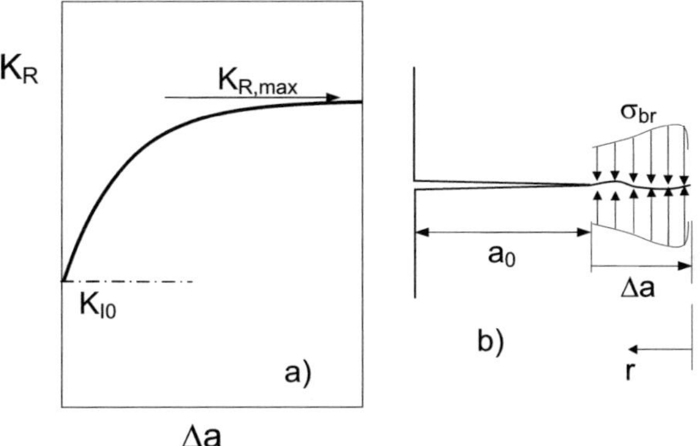

Fig. E7.1 a) Schematic of an increasing crack growth resistance curve starting from the crack-tip toughness K_{I0} with a rather linear increase and exhibiting a saturation value $K_{R,max}$, b) a crack in a ceramic material exhibiting crack surface interactions by bridging stresses.

A procedure that allows the bridging stresses to be determined from existing R-curve results was developed in [E7.4]. From the measured R-curves the bridging stress intensity factor K_{br} can be determined

$$K_R = K_{I0} - K_{br} , \quad K_{br} < 0 \tag{E7.1.1}$$

with the starting value K_{I0}, the so-called crack-tip toughness.

Using the weight function representation, the bridging stress intensity factor can be represented by the distribution of bridging stresses σ_{br} acting in the wake of the crack

$$K_{br}(\Delta a) = \int_{a_0}^{a_0+\Delta a} h(r,a)\,\sigma_{br}(\delta(r,a))\,dr \tag{E7.1.2}$$

with the fracture mechanics weight function h, the distance r from the tip, the initial crack length a_0 free of bridging, and the crack extension $\Delta a = a - a_0$. The bridging stresses depend on the actual crack opening displacements δ.

The total displacements in presence of bridging stresses result from superposition of the "bridging displacements" δ_{br} and the "applied displacements" δ_{appl} (the displacements under same load in the absence of the bridging stresses). It holds

$$\delta = \delta_{appl} + \delta_{br}$$

$$\delta_{br} = \frac{1}{E'} \int_{a-r}^{a} h(r,a')\,da' \int_{0}^{a'} h(r',a')\,\sigma_{br}(\delta(r',a))\,dr' \tag{E7.1.3}$$

with the plane strain modulus $E' = E/(1-v^2)$. The applied displacements are given by

$$\delta_{appl} = \frac{1}{E'} \int_{a-r}^{a} h(r,a')\,K_{appl}(a')\,da' \tag{E7.1.4}$$

with the applied stress intensity factor K_{appl} is given in fracture mechanics handbooks for various test specimens.

The system of equations (E7.1.2) and (E7.1.3) can be solved for instance by "successive approximation", starting with an in principle arbitrary first approximation of the bridging relation $\sigma_{br}(\delta)$. As the starting value for the displacements, the applied crack opening displacement field $\delta = \delta_{appl}$ may be used, resulting in the first distribution of the bridging stresses $\sigma_{br} = f(\delta_{appl}(r,a))$. These stresses have then to be introduced under the integral of (E7.1.3) yielding an improved distribution of the total displacements δ and, consequently, an improved bridging stress distribution, etc.

After a few iteration steps the solution of (E7.1.3) will converge so far the first approximation for the bridging law has not been chosen too unrealistic. This procedure has to be repeated for a number of crack lengths until the computed and measured R-curve values agree within a given limit.

E7.2 First- and second-order approximations

E7.2.1 First-order solution

It has to be emphasized that the procedures for solving the system of eqs.(E7.1.2) and (E7.1.3) are rather complicated and need much numerical effort. This was the reason for the development of first- and second-order approximations [E7.5, E7.6] which were successfully applied to silicon nitride [E7.7] and alumina ceramics.

This procedure consists in the following steps. By taking the derivative with respect to a on both sides of (E7.1.2) it results

$$\frac{dK_{br}}{da} = -\frac{dK_R}{da} = h(r,a)\sigma_{br}(r,a)\Big|_{r=\Delta a} + \int_{a_0}^{a}\frac{\partial h(r,a)}{\partial a}\sigma_{br}(r,a)\,dr$$

(E7.2.1)

$$+ \int_{a_0}^{a} h(r,a)\frac{\partial\sigma_{br}(r,a)}{\partial a}\,dr$$

The bridging stresses are then given as:

$$\sigma_{br}\Big|_{r=\Delta a} = -\frac{1}{h\big|_{r=\Delta a}}\frac{dK_R}{da} - \frac{1}{h\big|_{r=\Delta a}}\int_{a_0}^{a}\frac{\partial h(r,a)}{\partial a}\sigma_{br}(r,a,\Delta a)\,dr$$

(E7.2.2)

$$- \frac{1}{h\big|_{r=\Delta a}}\int_{a_0}^{a} h(r,a)\frac{\partial\sigma_{br}(r,a,\Delta a)}{\partial a}\,dr$$

Identifying $\sigma_{br}(r)=\sigma_{br}|_{r=\Delta a}$ gives a *first-order solution* for the bridging stresses:

$$\sigma_{br}^{(1)}(r) = -\frac{1}{h}\bigg|_{\Delta a=r}\frac{dK_R}{da}\bigg|_{\Delta a=r}$$

(E7.2.3)

The subscript $\Delta a=r$ in (E7.2.3) means: For the computation of the bridging stresses in a certain crack-tip distance r, e.g. $r=5$ μm, the derivative of the R-curve after $\Delta a=5$ μm has to be introduced as well as the value of the weight function h in distance $r=5$ μm.

E7.2.2 Second-order solution

In the first-order approximation only the first term of eq. (E7.2.2) was regarded with the bridging stresses depending exclusively on r. In a second-order solution, the first integral term of (E7.2.2) accounting for the influence of crack length via the weight function $h(r,a)$ may be included. In this approximation, the first-order solution (E7.2.3) has be used under the integral, resulting in

$$\sigma_{br}^{(2,1)}(r) = \sigma_{br}^{(1)}(r) - \frac{1}{h\big|_{r=\Delta a}} \int_{a_0}^{a} \frac{\partial h(r',a)}{\partial a} \sigma_{br}^{(1)}(r')\, dr' \qquad (E7.2.4)$$

where the first number in the superscript counts for the second term of (E7.2.2), and the second number for the step of iteration (started with $m=1$). Generally it holds for higher iteration numbers $m>1$

$$\sigma_{br}^{(2,m)}(r) = \sigma_{br}^{(1)}(r) - \frac{1}{h\big|_{r=\Delta a}} \int_{a_0}^{a} \frac{\partial h(r,a)}{\partial a} \sigma_{br}^{(2,m-1)}(r)\, dr \qquad (E7.2.5)$$

E7.2.3 An example of application

An R-curve for Mg-La-containing silicon nitride is shown in Fig. E7.2a. The crack propagation test was carried out on a pre-notched bending bar. The best fit of the measured data was given by

$$K_R = K_{I0} + C_1(1 - (1 + C_2\sqrt{\Delta a})\exp[-C_2\sqrt{\Delta a}]) \qquad (E7.2.6)$$

with the "best" set of coefficients

$$C_1 = 5.0 \text{ MPa}\sqrt{m}, \quad C_2 = 1542/\sqrt{m}, \quad K_{I0} = 2 \text{ MPa}\sqrt{m}$$

represented in Fig. E7.2a by the curve.

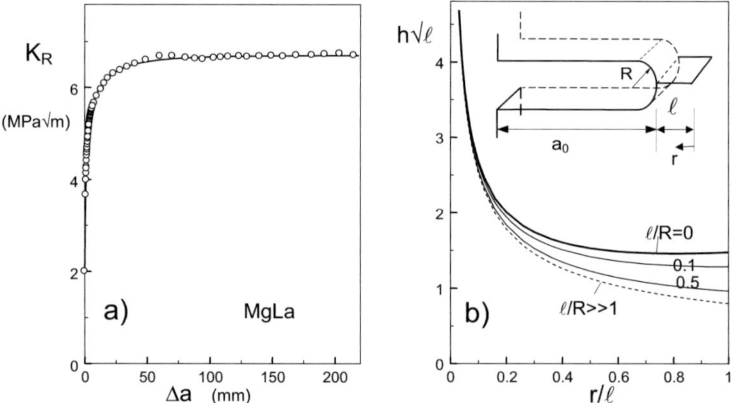

Fig. E7.2 a) R-curve for Mg-La-containing Si_3N_4, b) weight functions for a crack ahead of a slender notch.

For the determination of the bridging stresses, the weight function for a small crack ahead of a slender notch has to be applied. It is given by eqs.(E3.3.4a) and (E3.3.4b). Figure E7.2b shows this solution together with the geometric data R, ℓ, and r. For very short crack extensions in the order of about $\ell \leq R$, the weight function significantly depends on ℓ/R. Application of eq. (E7.2.3) yields the first-order bridging stresses as represented in Fig. E7.3a by the dash-dotted curve. The first and second iterative solutions of eq. (E7.2.5) are shown by the continuous curves. The third and fourth iterations showed differences less than 2 MPa, therefore, they are represented by the same dashed curve. This curve indicates the final curve of convergence, i.e. the bridging stress distribution $\sigma_{br}(r)$.

The full solution obtained by solving the simultaneous system of integral equations numerically, eqs.(E7.1.2) and (E7.1.3), and fitted by

$$\sigma_{br} \cong \sum_{n=0}^{N} \sigma_n \frac{\delta}{\delta_n} \exp[-\delta/\delta_n] \qquad (E7.2.7)$$

with $N=1$, σ_0=-3375 MPa, δ_0=0.0118μm, σ_1=-450 MPa, δ_1=0.0476μm, is shown in Fig. E7.3b as the solid curve. Although differences to the approximations occur, the second-order solution clearly exhibits all characteristic features of the exact solution, which are in this case:

a) Very high bridging stresses of $\sigma_{br}\approx$-1250 MPa,

b) a strong concentration of the bridging effects in a crack tip distance of about r=0-10μm.

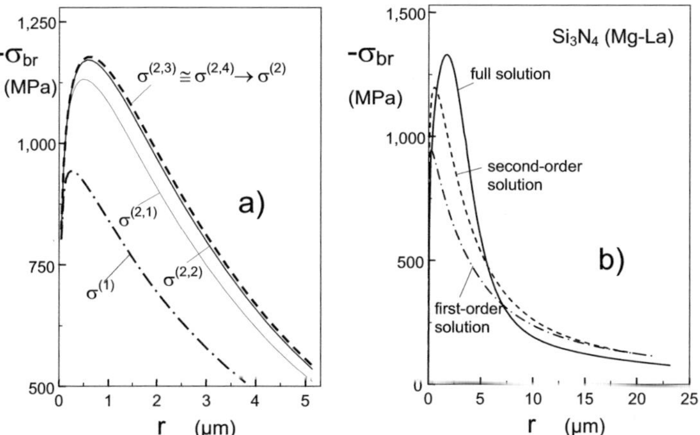

Fig. E7.3 a) Convergence study of the approximations; dash-dotted curve: first-order solution from (E7.2.3), continuous curves: first and second iteration steps for the second-order solution, dashed curve: identical third and fourth iteration steps indicate the final second-order solution, b) first- and second-order bridging stresses compared with the full solution obtained by simultaneous solving the integral equations (E7.1.2) and (E7.1.3).

The reason for the differences between the approximations and the full solution becomes evident by the representations in Fig. E7.4. Figure E7.4a shows the total crack profiles for a long crack extension of $\Delta a=50\mu m$ as the thick curve. For shorter crack-growth phases of $\Delta a=2\mu m$, $5\mu m$, $10\mu m$ and $20\mu m$, the total displacements were additionally computed via eq.(E7.1.3). The results are given by the thin curves with the circles indicating the individual crack length increments.

For a very short crack extension of $\Delta a=2\mu m$, the total displacements differ maximum by about 37% from the long-crack displacements. Consequently, also the bridging stresses must differ for such a small amount of crack propagation.

The related bridging stress distributions are given in Fig. E7.4b exhibiting clearly the differences in the stresses. These differences cause the different stress distributions for the full solution and the second-order approximation in Fig. 7.3b. In this context it should be taken into account that the stress values resulting from eqs. (E7.2.3) and (E7.2.5) are those acting at $r=\Delta a$, i.e. at the end of the bridging zone. These values are indicated by the circles in Fig. 7.4b. The dashed curve which interconnects these data points exhibits the main features of the second-order solution as slightly lower stress level and peak stress at a shorter crack-tip distance.

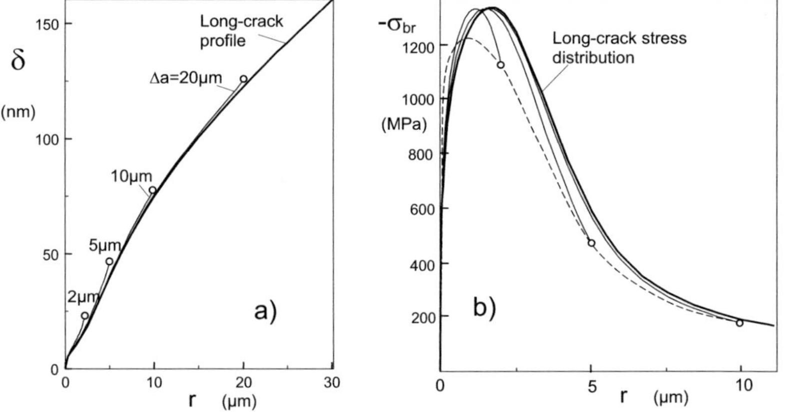

Fig. E7.4 a) Total displacement profiles for short crack extension phases, b) related deviations of the bridging stress distributions, dashed curve tentatively introduced as the interpolation of the stresses at $r=\Delta a$, represented by the circles.

E7.3 Displacements and weight functions

E7.3.1 Derivation of the weight function integral for displacements

The basis of the interrelation between displacements δ and weight function h is the weight function equation by Rice [E7.8]. For a straight crack it holds generally

$$h = \frac{E'}{K}\frac{\partial \delta}{\partial a} \qquad \text{(E7.3.1)}$$

Here it has to be emphasized that the derivative in (E7.3.1) is only a partial variation of the displacements for a fixed loading, i.e. for a fixed stress distribution. As outlined in [E7.9] integration of (E7.3.1) gives

$$\delta = \frac{1}{E'} \int\limits_{a-r}^{a} h(r,a')da' \int\limits_{0}^{a'} h(r',a')\sigma(r',a)dr' \qquad \text{(E7.3.2)}$$

where r is the distance from the crack tip. As the consequence of the partial derivative, in the inner integral, the stress is $\sigma(r',a)$ and <u>not</u> $\sigma(r',a')$. This means that the stress for the real crack length a has to be kept constant during integration over virtual crack increments. This fact is especially important for cases in which the stresses for differently long cracks become different for instance in the case of materials exhibiting bridging stresses.

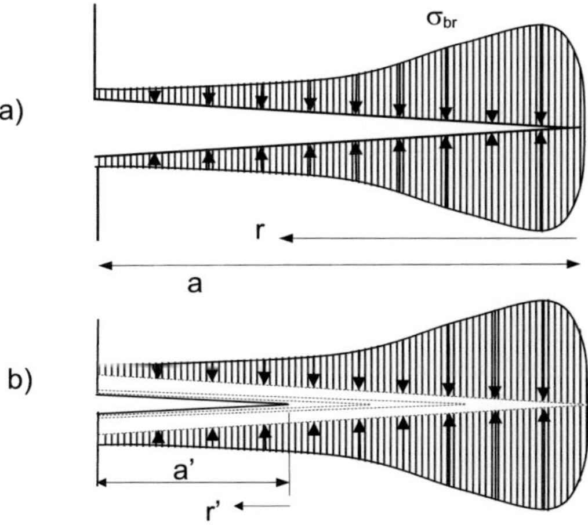

Fig. E7.5 Interpretation of the integral eq.(E7.3.2): a) bridging stress distribution in a real crack of length a, b) extension of a virtual crack of length $a'=a-r$ to $a'=a$ affected by the stress field in a).

The interpretation of the integrations in (E7.3.2) may be illustrated in Fig. E7.5. A crack of length a is shown under a bridging stress distribution $\sigma=\sigma_{br}(r,a)$. The displacements at the crack-tip distance r have to be computed (Fig. E7.5a). In this fixed stress field, a crack extends virtually from $a'=a-r$ (Fig. E7.5b) to $a'=a$. This *virtual* crack extension should not be misunderstood as a *real* crack propagation.

E7.3.2 Application to a Vickers indentation test on a ceramic material

E7.3.2.1 Solution of the system of integral equations

As an example, a Vickers indentation crack (Fig. E7.6a) is considered in a material that exhibits bridging behaviour. The actually present applied stress intensity factor K_{appl} after removing the indenter is related to the applied displacements δ_{appl}. An analytical solution for the COD of Vickers indentation cracks was given in [E7.10] (see Section F7.3).

Figure E7.6b shows displacement measurements on a silicon nitride ceramic containing MgO and Y_2O_3 (denoted as MgY) in the boundary glass phase. The bridging displacements δ_{br} caused by the bridging stresses σ_{br} result from

$$\delta_{br}(r) = \frac{1}{E'}\int_r^a\left(\int_0^{a'}h(r',a')\sigma_{br}(r')\,dr'\right)h(r,a)\,da' = \delta - \delta_{appl} \qquad (E7.3.3)$$

where δ stands for the measurable total displacements (for $r=a-x$, see Fig. E7.6a). The applied stress intensity factor K_{appl} has to be computed from

$$K_{appl} = K_{10} - \int_0^a h(r,a)\sigma_{br}(r)\,dr \qquad (E7.3.4)$$

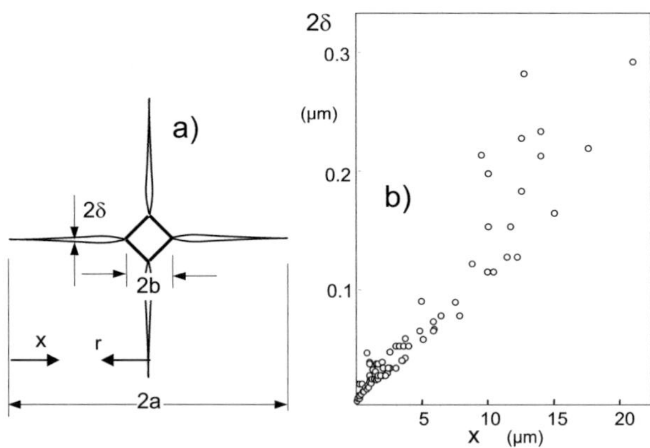

Fig. E7.6 a) Vickers indentation cracks (geometric parameters), b) crack opening displacement measurements for a Vickers indentation crack.

In (E7.3.4) the integral term is the bridging stress intensity factor. The quantity h in (E7.3.4) and (E7.3.3) is the fracture mechanics weight function. The weight function for the semi-circular crack can be obtained from eq.(F1.4)

The bridging stresses can now be determined by solving the integral equation (E7.3.3), simultaneously with eq.(E7.3.4). For this purpose we used the standard procedure for such problems which is called the "method of successive approximation".

From the evaluation of edge cracks a bridging relation of the type eq.(E7.2.7) was found. Since this relation should be independent of the special crack type, we used a set-up of the form eq.(E7.2.7) with free parameters σ_n and δ_n.

This bridging stress relation has to be inserted into eqs.(E7.3.3) and (E7.3.4). As the starting values, the coefficients $\sigma_n=0$ were used. Trivially, the first approximations are $K_{appl}=K_{I0}$ and $\delta=\delta_{appl}$. As a further free parameter an arbitrary value of K_{I0} has to be introduced. Use of $\delta=\delta_{appl}$ gives the first approximation of the bridging stresses $\sigma_{br}(r)$ from (E7.2.7). These stresses are then introduced in (E7.3.3) resulting in improved displacements $\delta(r)$. The next higher approximation is obtained by inserting the new $\delta(r)$ in (E7.2.7), etc. Convergence of the computed results indicates the solution for the arbitrarily given set of free parameters. In an "outer loop" of the iterative procedure, a systematic variation of the free parameters σ_n, δ_n and K_{I0} has to be performed. The best fit of the computed and measured total displacements then provides the solution of the system of equations with the best set of parameters.

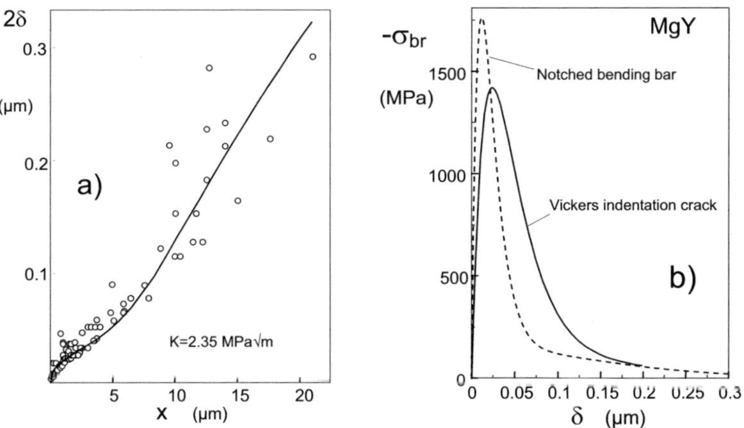

Fig. E7.7 Crack opening data of Fig. E7.6b (symbols) and "best" total COD solution (curve), b) related bridging stress relation (solid curve) and result from a notched bending bar (dashed curve).

The best approximation to the measured data in Fig. E7.7a is introduced by the curve. The related bridging law is shown in Fig. E7.7b as the solid curve. Additionally, the bridging relation from a test with a notched bending bar is represented by the dashed curve [E7.11]. The Vickers curve shows an about 20% lower peak stress and an increased "width". In this context it has to be noted that the data scatter in the R-curves (from which the dashed curve was derived) are clearly less those of the COD measurements. From the described evaluation also the crack-tip toughness was obtained, with the result of $K_{I0} \cong 2.3$ MPa√m.

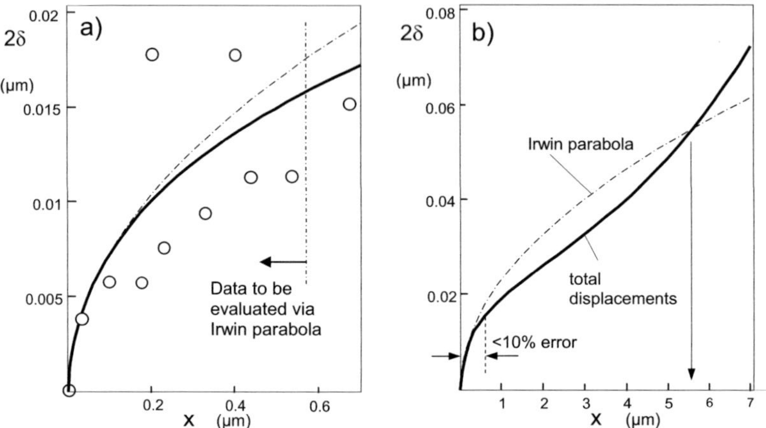

Fig. E7.8 Comparison of total near-tip displacements (solid curve) with the Irwin parabola (dash-dotted curve).

E7.3.2.2 Near-tip COD-behaviour

The near-tip results of Fig. E7.7a are plotted in Fig. E7.8a with an increased resolution. In addition to this solution (solid curve) also the near-tip solution given by the Irwin parabola

$$\delta_0 \overset{x \to 0}{=} \sqrt{\frac{8}{\pi}} \frac{K_{I0}}{E} \sqrt{x} \tag{E7.3.5}$$

(K_{I0}=crack-tip toughness) is plotted (dash-dotted curve). These two curves agree in a very small crack-tip distance as theory requires. For larger crack-tip distances clear deviations can be observed (Fig. E7.8b). In the region 0.6 µm< x < 5.5 µm the Irwin parabola overestimates the total displacements and for all larger distances, x > 5.5 µm, it underestimates the total COD.

Tolerating errors less than 10% we can conclude that for the given material the application of the Irwin parabola is restricted to distances of x < 0.6 µm.

E7.4 The effect of the rising R-curve on the compliance

In Section E3 the crack length evaluation via compliance was addressed for the case of free crack faces. In order to include the influence of the bridging interactions on the compliance, a modified relation may be used. Application of the compliance procedure needs a complete computation of the bridging-influenced compliance.

By using Betti's theorem for the reciprocal works, the displacements at the loading points caused by the bridging effect, $\delta_{LP,br}$, result as

$$\tfrac{1}{2B} P_{appl}\delta_{LP,br} = \int_0^{\Delta a}\sigma_{br}(r)\delta_{appl}(r)\,dr \tag{E7.4.1}$$

with the factor $1/(2B)$ since the total force P_{appl} in a 4-point bending is split into <u>two</u> line loads over thickness B at the two loading points. The origin of the coordinate r is chosen to be located at the crack tip. The "applied" crack-opening displacements are

$$\delta_{appl} = \frac{1}{E'}\int_{a-r}^{a} h(r,a')K_{appl}(a')\,da' \tag{E7.4.2}$$

with the applied stress intensity factor K_{appl} and the weight function h.

In principle, the solution of eq.(E7.4.1) may be found iteratively:

- In the first approximation, the bridging effect on the compliance is ignored. With this approximation of the R-curve the bridging law has to be determined by solving the simultaneous integral equations (E7.1.2) and (E7.1.3). The loading point displacements due to the bridging stresses are then known from (E7.4.1).

- The new compliance including the loading point displacements due to bridgingyields a new crack length and, consequently, an improved R-curve. Evaluation of this improved solution in the same way converges to the correct crack lengths and, finally, the correct R-curve.

The delineated procedure needs extremely much effort and may, therefore, be replaced by an approximation. As a first-order solution for the bridging stresses σ_{br} in the crack wake it results as shown in Section E7.2

$$\sigma_{br}(r) \cong -\frac{1}{h}\frac{dK_R}{da}\bigg|_{\Delta a=r} \tag{E7.4.3}$$

The total displacements by action of externally applied and intrinsic bridging stresses simultaneously, ΔC, are then given in terms of the applied and bridging compliances

$$\Delta C \cong \Delta C_{appl}(\Delta a) - \frac{2B}{P_{appl}^2}\int_0^{\Delta a}\frac{\delta_{appl}}{h}\frac{dK_R}{da}\bigg|_{\Delta a=r}dr \tag{E7.4.4}$$

Whereas the left-hand side is a result from measurements, the unknown true crack increment Δa can be obtained form the right side of (E7.4.4).

The unknown increment Δa appears explicitly at two places in (E7.4.4), namely in the argument of the compliance function for the applied load and at the upper integration limit of the bridging part. Implicitly, the crack length also affects the K_R-term of the integrand. This fact calls for an iterative solution using the method of "successive approximation". For its explanation, eq.(E7.4.4) may be slightly re-written as

$$\Delta C \cong \Delta C_{appl}(\Delta a^{(n+1)}) - \frac{2B}{P_{appl}^2} \int_0^{\Delta a^{(n+1)}} \left(\frac{\delta_{appl}}{h} \frac{dK_R}{da}\bigg|_{\Delta a=r} \right)^{(n)} dr \qquad (\text{E7.4.5})$$

In the first approximation step, the compliance in the absence of bridging effects is used to compute the crack depth $a^{(1)}$. From this, the first-order approximations for the applied displacements, the weight function and the R-curve K_R have to be computed as indicated by the superscript (n), here $n=1$. Using the first-order terms under the integral, eq.(E7.4.5) can be solved by application of a zero routine providing the second-order approximation $a^{(2)}$. The procedure has to be repeated until a certain degree of convergence is reached. This establishes the final crack depth increment Δa.

References E7

E7.1 Fett, T., Stress intensity factors, T-stresses, Weight functions, IKM 50, Universitätsverlag Karlsruhe, 2008.

E7.2 Fett, T., Determination of stresses from stress intensity factor solutions: An inverse fracture mechanics problem, Int. J. Fract. **70**(1995), R69-R75

E7.3 Munz, D., What can we learn from R-curve measurements? J. Am. Ceram. Soc. **90**(2007), 1-15.

E7.4 Fett, T., Munz, D., Thun, G., Bahr, H.A., Evaluation of bridging parameters in Al_2O_3 from R-curves by use of the fracture mechanical weight function, J. Amer. Ceram. Soc. **78**(1995) 949-51.

E7.5 Fünfschilling, S., Mikrostrukturelle Einflüsse auf das R-Kurvenverhalten bei Siliciumnitridkeramiken, Thesis, IKM 53, Universitätsverlag Karlsruhe, 2009.

E7.6 Fünfschilling, S., Fett, T., Gallops, S., Kruzic, J.J., Hoffmann, M.J., Oberacker, R., First- and second-order approaches for the direct determination of bridging stresses from R-curves, to be published in J. Europ. Ceram. Soc.

E7.7 Fünfschilling, S., Determination of R-curves from corrected load-displacement data and the calculation of the bridging stresses from the R-curves, 6th International Conference on Nitrides and Related Materials (ISNT 2009), 15.-18.03.2009, Karlsruhe.

E7.8 Rice, J.R., Some remarks on elastic crack-tip stress fields, Int. J. Solids and Structures **8**(1972), 751-758.

E7.9 Fett, T., Munz, D., Stress intensity factors and weight functions, Computational Mechanics Publications, Southampton, 1997.

E7.10 Fett, T., Kounga Njiwa, A.B., Rödel, J., Crack opening displacements of Vickers indentation cracks, Engng. Fract. Mech. **72**(2005), 647-659.

E7.11 Fünfschilling, S., Fett, T., unpublished results.

PART F

TWO-DIMENSIONAL CRACKS

Part F deals with:

Approximate weight function for half-penny-shaped surface cracks

Local stress intensity factors for semi-elliptical surface cracks, singularity problems

Averaged stress intensity factors, definition

Averaged stress intensity factors for non-continuous crack-face loading

Rectangular surface crack

Semi-elliptical cracks ahead of narrow notches

Circular and ring-shaped cracks under non-radial symmetric stresses

F1

Weight function for semi-circular cracks

A semi-circular surface crack is shown in Fig. F1.1a. For the case of a stress distribution depending on the radial coordinate r exclusively, the stress intensity factor along the crack front can be expressed in terms of a weight function by

$$K(\varphi) = \int_0^a h(r,\varphi)\,\sigma(r)\,dr \qquad (F1.1)$$

The weight function for the semi-circular crack can be obtained by matching h to the reference stress intensity factor for constant stress (see Fett and Munz, [F1.1]). Basis of this procedure may be the weight function for a circular crack in an infinite body, given as

$$h_{circ} = \frac{2r}{\sqrt{\pi a(a^2 - r^2)}} \qquad (F1.2)$$

For the semi-circular crack in the half-space the stress intensity factor solution for constant stress is known for instance from the tensile solution by Newman and Raju [F1.2]

$$K = \sigma_0 \sqrt{a\pi}\, Y, \quad Y = 1.04 \frac{2}{\pi}[1 + 0.1(1 - \sin\varphi)^2] \qquad (F1.3)$$

In this context it has to be considered that the simplest set-up of $h = h_{circ}\, K/K_{circ}$ is not correct since now also the singular term of the weight function would be affected. This part of course must remain unaffected. Therefore, only the non-singular part has to be modified. Adjusting the non-singular weight function part to the reference solution yields

$$h(r,\varphi) = \frac{2r[1 + c(1 - r/a)]}{\sqrt{\pi a(a^2 - r^2)}} \qquad (F1.4)$$

where the coefficient c accounts for the influence of the free surface

$$c = \frac{0.04 + 0.104(1 - \sin\varphi)^2}{1 - \frac{\pi}{4}} \qquad (F1.5)$$

The weight functions $h(r,\varphi)$ are plotted in Fig. F1.1b for $\varphi = 0$ and $\varphi = 90°$ as the thin lines. The influence of the angle φ is rather small. Consequently, it is recommended to neglect the angular influence by using an average value of $c \cong 0.42$. This value results in the thick curve. The dash-dotted curve represents the solution for the embedded circular crack. In Fig. F1.2 the weight function for the semi-circular crack is compared with those for edge-cracks of different relative crack depths a/W (W=specimen thickness). The main difference between the two crack types is the fact that at $r/a \rightarrow 0$ the weight function for the semi-circular crack disappears, whereas the edge crack shows finite values at $x/a \rightarrow 0$.

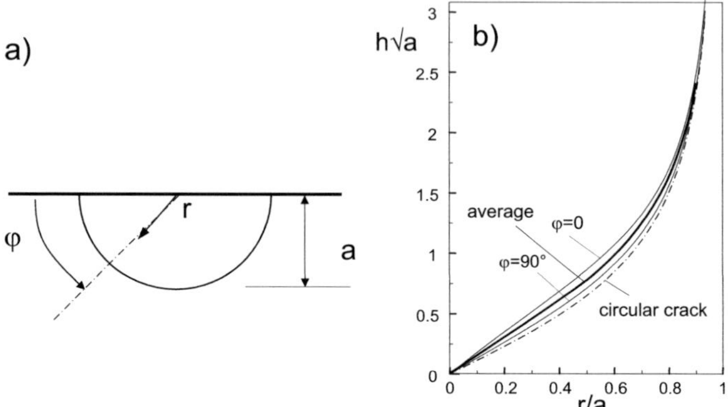

Fig. F1.1 a) Semi-circular surface crack, b) weight function for the semi-circular surface crack parallel to the surface ($\varphi = 0$) and normal to the surface (φ=90°) compared with the circular crack in a semi-infinite body.

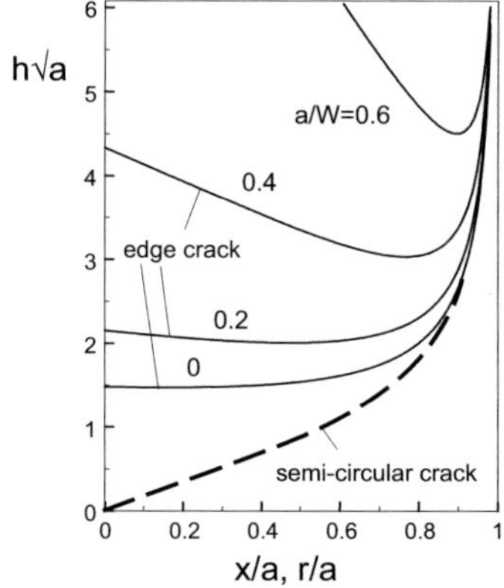

Fig. F1.2 Weight functions for differently deep edge cracks (solid curves) and for the semi-circular crack (dashed curve), W=specimen thickness.

References F1

F1.1 Fett, T., Munz, D., Stress intensity factors and weight functions, Computational Mechanics Publications, Southampton, 1997.
F1.2 Newman, J.C., Raju, I.S., An empirical stress intensity factor equation for the surface crack, Engng. Fract. Mech. **15**(1981) 185-192.

F2

Problems with local stress intensity factors at semi-elliptical surface cracks

F2.1 Local stress intensity factors

The well established relation of Newman and Raju [F2.1] for the stress intensity factor of a single semi-elliptical crack of depth a and width $2c$ in the half-space under remote stresses σ_0 may be used here. This relation reads

$$K = \sigma_0 \sqrt{a\pi}\, F$$

$$F = \frac{(1.13 - 0.09\frac{a}{c})}{\mathbf{E}(1 - a^2/c^2)}[1 + 0.1(1 - \sin\varphi)^2]\left[\left(\frac{a}{c}\right)^2 \cos^2\varphi + \sin^2\varphi\right]^{1/4} \qquad (\text{F2.1.1})$$

where $\mathbf{E}$ is the complete elliptical integral of second kind and φ the parametric ellipse angle for a location at the crack front (for φ see Fig. F3.6). The accuracy of eq.(F2.1.1) is very good. This can be seen for example by comparison with the highly-precise data reported by Isida et al. [F2.2] having a maximum error of 0.1%. Figure F2.1a shows the solution by Isida et al as the symbols and eq.(F2.1.1) as the curves. Maximum deviations between the two solutions are less than 2%. In Fig. 2.1b, curves for the region of $0.7 \leq a/c \leq 0.9$ are represented. That aspect ratio for which the stress intensity factors at points $\varphi = 0$ (commonly denoted as point B) and $\varphi = 90°$ (point A) are identical (squares in Fig. F2.1b), namely for $a/c = 0.8264$, is indicated by the dashed curve.

This diagram implies that:

- There exists no aspect ratio for a semi-ellipse that yields a constant stress intensity factor along the crack front.

- Even for the aspect ratio of $a/c = 0.8264$ for which the local stress intensity factors at the surface and the deepest point of the semi-ellipse are identical, the stress intensity factor shows a significant variation with the angle φ. This fact causes problems in describing crack propagation. Problems are obvious even in the simplest case of crack extension, namely, stable crack growth at $K = K_{Ic}$. By load application to a semi-elliptic crack of $a/c = 0.8264$, the critical stress intensity factor is at first reached at the surface and the deepest points. These regions can extend but the regions at which $K(\varphi) < K(0°, 90°)$ must stay behind a semi-ellipse (Fig. F2.1c). Already after the first infinitesimally small step of crack extension, the crack geometry must deviate from a semi-ellipse and, consequently, the semi-ellipse solution can no longer be applied to this now slightly irregular crack shape. This makes evident that crack growth prediction on the basis of local stress intensity factors is not simply possible.

A way out of this dilemma was proposed very early by Cruse and Besuner [F2.3]. From observation of the crack shape during stable and fatigue crack propagation it could be concluded that cracks propagated as semi-ellipses although local stress intensity factors did not allow this. An irregular shape to be expected from the local variation of the stress intensity factors was not detectable.

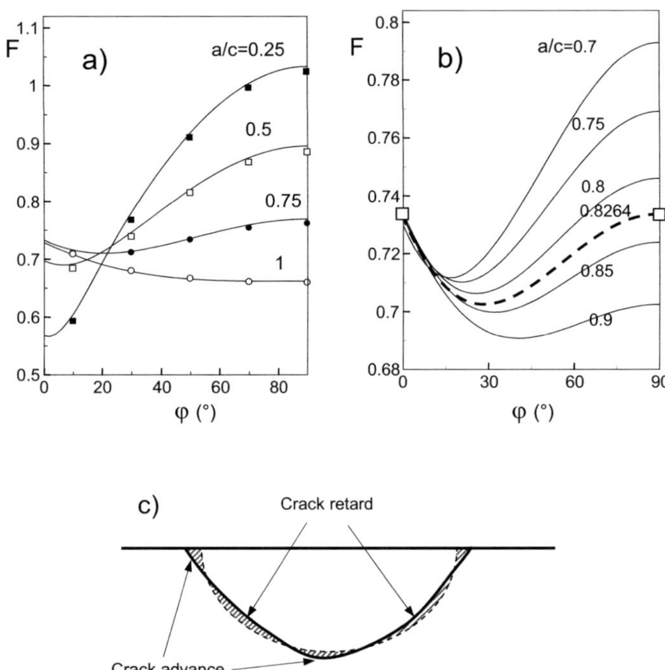

Fig. F2.1 a) Comparison of the relation by Newman and Raju [F2.1] (curves) with highly precise data by Isida et al.[F2.2] (symbols); b) the Newman-Raju solution for an aspect ratio range relevant for crack propagation in tension (dashed curve: special case a/c=0.8264 showing identical stress intensity factors at the surface and the deepest point), c) schematic of crack propagation for the semi-ellipse (dashed crack contour) with a/c=0.8264.

F2.2 Singularity behaviour at crack/surface intersections

A further effect that makes the applicability of local stress intensity factors problematic is the change of the stress singularity near the surface [F2.4, F2.5]. Figure F2.2a shows the top view of a surface breaking crack with an arbitrarily inclined crack front within the crack-plane (defining the angle φ).

The problem of a crack-front intersecting a free surface is well established in theoretical fracture mechanics and studied numerically by the "asymptotic analysis" [F2.6, F2.7]. As shown by Kondratiev [F2.8] the 3-dimensional displacement field near the intersection is asymptotically given by

$$\mathbf{u} = \begin{pmatrix} u_x \\ u_y \\ u_z \end{pmatrix} = \sum_i k_i r^{\lambda_i} \mathbf{U}_i ,$$
(F2.2.1)

where $\mathbf{u}$ is the vector of the 3 displacement components, r is the distance from the intersection point, $\mathbf{U}_i$ are angular functions for the three displacement components, k_i the so-called "corner intensity factors", and λ_i are the corner singularity exponents.

For cracks, up to 4 singularity exponents can exist. Under pure symmetrial conditions only the *symmetrical solution* occurs

$$\mathbf{u} = \begin{pmatrix} u_x \\ u_y \\ u_z \end{pmatrix} = k\, r^{\lambda} \mathbf{U} .$$
(F2.2.2)

Figure F2.2b shows the singularity exponent λ for the symmetric solution as a function of the terminating angle φ. It should be noted that $0 \le \lambda \le 1$ results in all cases. The amplitude of the asymptotic field is proportional to the stress intensity factor present in larger distance from the surface.

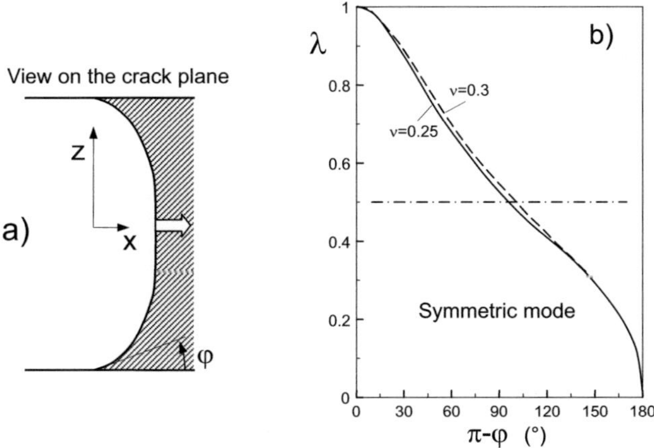

Fig. F2.2 a) Geometrical data for a crack terminating at free surfaces, b) singularity exponent for the symmetrical mode [F2.6].

With the abbreviation $\mu=1-\varphi/\pi$, the singularity exponent for $\nu=0.25$ can be approximated by

$$\lambda \cong 1 - 6.564\mu^2 + 14.4255\mu^3 - 9.342\mu^4 - 3.888\mu^5 + 4.3684\mu^6 + 0.972\mu^7 \sqrt{1-\mu} \qquad \text{(F2.2.3)}$$

From the displacements u the stresses σ can be derived with the general result of

$$\sigma \propto r^{\lambda-1} \qquad \text{(F2.2.4)}$$

In the special case of a crack terminating perpendiculary to the free surface ($\varphi=90°$) as ocurring for semi-elliptical surface cracks, the singularity exponent is $\lambda=0.53$ and the stresses show a singularity of $\sigma \propto r^{-0.47}$, called a "weak singularity". Normal stress intensity factors describing $1/\sqrt{r}$ singularities are no longer applicable.

From curves reported in [F2.9] it can be concluded that $\lambda=1/2$ is only fulfilled for a crack terminating angle of

$$\varphi \cong 90° - 38.8°\,\nu \qquad \text{(F2.2.5)}$$

(ν=Poisson's ratio). For materials with $0.2 \leq \nu \leq 0.3$, the span of possible crack terminating angles is $78° \leq \varphi \leq 82°$, i.e. deviations of 8-12° from the normal.

References F2

F2.1 Newman, J.C., Raju, I.S., An empirical stress intensity factor equation for the surface crack, Engng. Fract. Mech. **15**(1981) 185-192.

F2.2 Isida, M., Noguchi, H., Yoshida, T., Yoshida, T., Tension and bending of finite thickness plates with a semi-elliptical surface crack, Int. J. Fract. **26**(1984), 157-188.

F2.3 Cruse, T.A., Besuner, P.M., Residual life prediction for surface cracks in complex structural details, J. of Aircraft **12**(1975), 369-375.

F2.4 Benthem, J.P., "State of stress at the vertex of a quarter-infinite crack in a half-space," *Int. J. Solids and Struct.* **13**(1977),479.

F2.5 Bazant, Z.P., Estenssoro, L.F., "Surface singularity and crack propagation," *Int. J. Solids and Struct.*, **15**(1979), 405-426.

F2.6 Dimitrov, A., Buchholz, F.-G., Schnack, E., "3D-corner effects in crack propagation." In H.A. Mang, F.G. Rammerstorfer, and J. Eberhardsteiner, editors, *On-line Proc. of the 5th World Congress in Comp. Mech. (WCCMV)*, Vienna, Austria, July 7–12 2002. http://wccm.tuwien.ac.at/index.html.

F2.7 Dimitrov, A., Buchholz, F.-G., Schnack, E., "On three-dimensional effects in propagation of surface-breaking cracks, CMES, Comp. Mod. in Engng.& Science **12**(2006),1-25.

F2.8 Kondratiev, V., "Boundary problems for elliptic equations in domains with conical or angular points," *Transact. Moscow Math. Soc.* **16**(1967)**, 227-313.

F2.9 Dimitrov, A., Eckensingularitäten bei räumlichen Problemen der Elastizitätstheorie: Numerische Berechnung und Anwendungen, PhD-Thesis, University of Karlsruhe, 2002.

F3

Application of average stress intensity factors

F3.1 Weight function approach

F3.1.1 Definition of stress intensity factor

The weight function procedure applicable to any 2-dimensional crack problem is based on the relation of Rice [F3.1], which relates the variation of the crack opening displacement v_r in a certain reference loading case, e.g. σ_r = const., to the stress intensity factors in the actual loading case σ

$$\frac{1}{\Delta S}\int_{(\Delta S)} K\,K_r\,d(\Delta S) = E'\int_{(S)} \sigma\frac{\partial v_r}{\partial(\Delta S)}\,dS \qquad (\text{F3.1.1})$$

with K_r = reference stress intensity factor, $E' = E/(1-v^2)$, E = Young's modulus, v = Poisson's ratio, and $d(\Delta S)=dL\times\delta\ell$ (see Fig. F3.1).

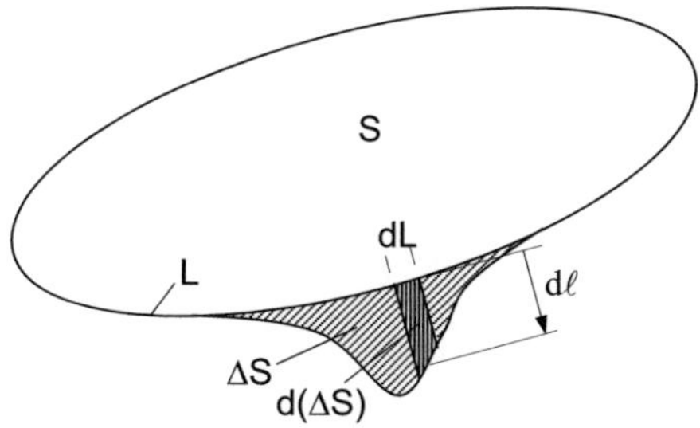

Fig. F3.1 Crack of area S showing an arbitrary virtual in-plane crack area increment ΔS.

A rather rough derivation of this relation can be given (see [F3.2]):
Two loading cases (subscripts 1 and 2) are considered. The energy release rates for a virtual crack extension ΔS are given by

$$G_1 = \frac{\partial}{\partial(\Delta S)}\int_{(S)} v_1\sigma_1\,dS = \frac{1}{E'\Delta S}\int_{(\Delta S)} K_1^2\,d(\Delta S) \qquad (\text{F3.1.2})$$

$$G_2 = \frac{\partial}{\partial(\Delta S)}\int_{(S)} v_2\sigma_2\,dS = \frac{1}{E'\Delta S}\int_{(\Delta S)} K_2^2\,d(\Delta S) \qquad (\text{F3.1.3})$$

For the combined load $\sigma_3=\sigma_2+\sigma_1$ it results trivially

$$G_3 = \frac{\partial}{\partial(\Delta S)} \int\limits_{(S)} (v_1 + v_2)(\sigma_1 + \sigma_2)dS = \frac{1}{E'\Delta S} \int\limits_{(\Delta S)} (K_1 + K_2)^2 d(\Delta S) \qquad (F3.1.4)$$

After multiplying the expressions in brackets and introducing (F3.1.2) and (F3.1.3)

$$G_3 = \frac{\partial}{\partial(\Delta S)} \int\limits_{(S)} (v_1\sigma_2 + v_2\sigma_1)dS = \frac{2}{E'\Delta S} \int\limits_{(\Delta S)} K_1 K_2 d(\Delta S) \qquad (F3.1.5)$$

By making use of Betti's theorem

$$\int\limits_{(S)} v_1\sigma_2 dS = \int\limits_{(S)} v_2\sigma_1 dS \qquad (F3.1.6)$$

and taking the derivative under the integral, it finally results

$$\int\limits_{(S)} \sigma_2 \frac{\partial v_1}{\partial(\Delta S)} dS = \frac{1}{E'\Delta S} \int\limits_{\Delta S} K_1 K_2 d(\Delta S) \qquad (F3.1.7)$$

If we identify the reference loading case (subscript r) with the loading case "1" and the actual load case with "2" the weight function equation reads

$$\int\limits_{(S)} \sigma \frac{\partial v_r}{\partial(\Delta S)} dS = \frac{1}{E'\Delta S} \int\limits_{\Delta S} K K_r d(\Delta S) \qquad (F3.1.8)$$

This relation is identical with (F3.1.1). A procedure for the determination of the crack opening displacement field v_r for a semi-elliptic surface crack under a constant reference stress was given in [F3.3]. A disadvantage of the weight function relation (F3.1.1) is the circumstance that the local stress intensity factor K occurs under integral sign and, consequently, cannot be isolated explicitly. This fact gives rise for the definition of a so-called "averaged stress intensity factor" $\overline{K}$.
This stress intensity factor is *defined* by the left side of eq.(F3.1.1) as

$$\overline{K} \stackrel{def}{=} \frac{\dfrac{1}{\Delta S} \int\limits_{(\Delta S)} K K_r d(\Delta S)}{\sqrt{\dfrac{1}{\Delta S} \int\limits_{(\Delta S)} K_r^2 d(\Delta S)}} \qquad (F3.1.9)$$

The denominator in this definition ensures self-consistency of the so-defined $\overline{K}$. If the actual loading case and the reference loading case (mostly chosen as remote tension) are identical, $\sigma = \sigma_r$ (resulting in $K = K_r$), equation (F3.1.9) reads

$$\overline{K}_r = \frac{\dfrac{1}{\Delta S} \int\limits_{(\Delta S)} K_r^2 d(\Delta S)}{\sqrt{\dfrac{1}{\Delta S} \int\limits_{(\Delta S)} K_r^2 d(\Delta S)}} = \sqrt{\frac{1}{\Delta S} \int\limits_{(\Delta S)} K_r^2 d(\Delta S)} \qquad (F3.1.10)$$

Using this result, the definition eq.(F3.1.9) can be expressed in the form

$$\overline{K} \stackrel{def}{=} \frac{\dfrac{1}{\Delta S} \displaystyle\int_{(\Delta S)} K K_r \, d(\Delta S)}{\overline{K}_r} \tag{F3.1.11}$$

Consequently, (F3.1.1) can be written in the usual form

$$\overline{K} = \frac{E'}{\overline{K}_r} \int_{(S)} \sigma \frac{\partial v_r}{\partial(\Delta S)} \, dS \tag{F3.1.12}$$

F3.1.2 Virtual crack area increments proposed by Cruse and Besuner

In the considerations of Section F3.1.1 no special type of the virtual crack area increment ΔS was assumed. For a numerical evaluation of course we have to make an appropriate choice. As a consequence of experimental observation of crack growth, Cruse and Besuner [F3.4] suggested two independent virtual crack changes ΔS (Fig. F3.2a) which preserve the semi-elliptical crack shape, namely, crack depth increment Δa with width c kept constant

$$\Delta S_A = \tfrac{1}{2}\pi c \Delta a, \quad d(\Delta S_A) = c\Delta a \sin^2 \varphi \, d\varphi \tag{F3.1.13}$$

or crack width increment Δc with depth a=const

$$\Delta S_B = \tfrac{1}{2}\pi a \Delta c, \quad d(\Delta S_B) = a\Delta c \cos^2 \varphi \, d\varphi \tag{F3.1.14}$$

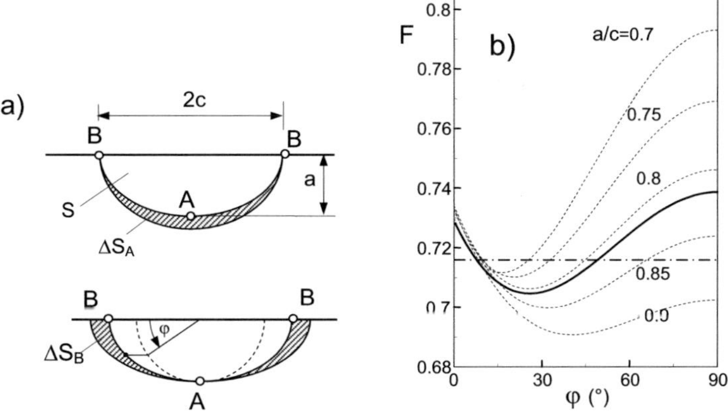

Fig. F3.2 a) Virtual crack extensions according to Cruse and Besuner [F3.4], b) equilibrium semi-ellipse under remote tension (solid curve) showing an aspect ratio of $(a/c)_{eq}$=0.812 and F_A=F_B=0.717 (horizontal dash-dotted line).

F3.1.3 Equilibrium semi-ellipse

That semi-ellipse simultaneously growing in depth and width directions is called the equilibrium semi-ellipse. The condition for this semi-ellipse with aspect ratio $(a/c)_{eq}$ reads

$$\overline{K}_A = \overline{K}_B \qquad (\text{F3.1.15})$$

or

$$F_A = F_B \qquad (\text{F3.1.16})$$

with the geometric functions F_A and F_B defined by

$$\overline{K}_{A,B} = \sigma_0 F_{A,B} \sqrt{\pi a} \qquad (\text{F3.1.17})$$

In terms of the suggestion by Cruse and Besuner [F3.4], this condition yields

$$\int_0^{\pi/2} K(\varphi) K_r(\varphi) \sin^2 \varphi \, d\varphi = \int_0^{\pi/2} K(\varphi) K_r(\varphi) \cos^2 \varphi \, d\varphi \qquad (\text{F3.1.18})$$

For the case of constant remote tension $\sigma = \sigma_r = \text{const}$, the evaluation of eq.(F3.1.18) gives $(a/c)_{eq} = 0.812$. The geometric functions for the local stress intensity factors for this geometry are shown in Fig. F3.2b by the solid curve, the geometric functions $F_A = F_B = 0.717$ are represented by dash-dotted horizontal line.

F3.2 Approximate interpretation of the average stress intensity factor

A fracture mechanics interpretation of the abstractly defined averaged stress intensity factors according to eqs.(F3.1.9) or (F3.1.11) was early tried by Cruse and Besuner [F3.4]. They related the averaged stress intensity factors to the energy release rate concept in the following way:

The mean energy release rate for the virtual crack extension ΔS, denoted as $G_{\Delta S}$ can be computed from the local values of the energy release rate

$$G_{\Delta S} = \frac{1}{\Delta S} \int_{(\Delta S)} G \, d(\Delta S) = \frac{\partial}{\partial(\Delta S)} \int_{(S)} v\sigma \, dS \qquad (\text{F3.2.1})$$

Using the relation between energy release rate and the local stress intensity factors

$$G = \frac{K^2}{E'} \qquad (\text{F3.2.2})$$

one obtains

$$G_{\Delta S} = \frac{1}{E' \Delta S} \int_{(\Delta S)} K^2 d(\Delta S) \qquad (\text{F3.2.3})$$

From $G_{\Delta S}$, Cruse and Besuner computed an average stress intensity factor K^*

$$K^* = \sqrt{G_{\Delta S} E'} \qquad (\text{F3.2.4})$$

A stress intensity factor defined in this way does not fulfil the weight function equation (F3.1.1) as had been shown in [F3.5]. It holds for K^*

$$K^* K_r^* \geq E' \int_{(S)} \sigma \frac{\partial v_r}{\partial(\Delta S)} dS \qquad (F3.2.5)$$

with the equality sign valid only in the reference loading case $\sigma = \sigma_r$, $K^* = K_r^*$.

The reason for the inequality sign in the general case of $\sigma \neq \sigma_r$ is the inequality of Schwarz for integrals which says that

$$\sqrt{\int_{(\Delta S)} K^2 d(\Delta S)} \bullet \sqrt{\int_{(\Delta S)} K_r^2 d(\Delta S)} \geq \int_{(\Delta S)} K K_r d(\Delta S) \qquad (F3.2.6)$$

In their basic investigations Cruse and Besuner used the equality sign for any loading case. Although this is not exact it could be shown that for practical purposes the differences between $\overline{K}$ and K^* are often negligible [F3.5].

F3.3 Average stress intensity factors for tensile and bending loading

F3.3.1 Stress intensity factors for remote tension

Geometric functions F_A and F_B for semi-elliptical cracks under remote tension are shown in Fig. F3.3a. For the ranges of $0.7 \leq a/c \leq 1$ and $0 \leq a/t \leq 0.15$, they read

$$F_A = 1.118 - 0.6127 \frac{a}{c} + 0.1609 \left(\frac{a}{c}\right)^2 + 0.8449 \left(\frac{a}{t}\right)^2 - 0.7257 \left(\frac{a}{c}\right)\left(\frac{a}{t}\right)^2 \qquad (F3.3.1)$$

$$F_B = 0.7683 - 0.00746 \frac{a}{c} - 0.0534 \left(\frac{a}{c}\right)^2 + 0.7967 \left(\frac{a}{t}\right)^2 - 0.5529 \left(\frac{a}{c}\right)\left(\frac{a}{t}\right)^2 \qquad (F3.3.2)$$

$$\frac{F_B}{F_A} \cong 0.6421 + 0.531 \frac{a}{c} - 0.111 \left(\frac{a}{c}\right)^2 + 0.0022 \left(\frac{a}{t}\right)^2 + 0.16 \left(\frac{a}{c}\right)\left(\frac{a}{t}\right)^2 \qquad (F3.3.3)$$

with the thickness t (in contrast to one-dimensional cracks for which the thickness is mostly abbreviated by W). As can be seen from Fig. F3.3, the influence of a/t representing the influence of the free rear wall is very small for pure tension.

The aspect ratio of the equilibrium ellipse depending on the crack depth is

$$\left(\frac{a}{c}\right)_{eq} \cong 0.812 - 0.350 \left(\frac{a}{t}\right)^2 \qquad (F3.3.4)$$

For the aspect ratio $(a/c)_{eq}$ of the equilibrium ellipse we obtain a maximum influence of less than 1% for $a/t \leq 0.15$. The relative variation of the stress intensity factors $F_A = F_B$ is plotted in Fig. F3.4. For $a/t \leq 0.15$, the variation of the stress intensity factors is less than 0.35% and may be neglected in practice.

The data for an extended region of $0.25 \leq a/c \leq 1$ can be approximated for $a/t=0$ by

$$F_A \cong 1.191 - 0.7876\frac{a}{c} + 0.2582\left(\frac{a}{c}\right)^2 \qquad (F3.3.5)$$

$$F_B \cong 0.699 + 0.1541\frac{a}{c} - 0.149\left(\frac{a}{c}\right)^2 \qquad (F3.3.6)$$

$$\frac{F_B}{F_A} \approx 0.632 + 0.553\frac{a}{c} - 0.121\left(\frac{a}{c}\right)^2 \qquad (F3.3.7)$$

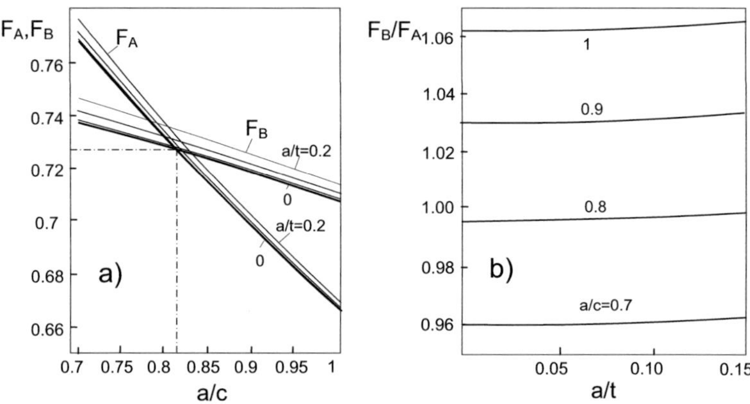

Fig. F3.3 Average stress intensity factors for a semi-elliptical crack under remote tension; a) geometric functions, b) ratio of the stress intensity factors.

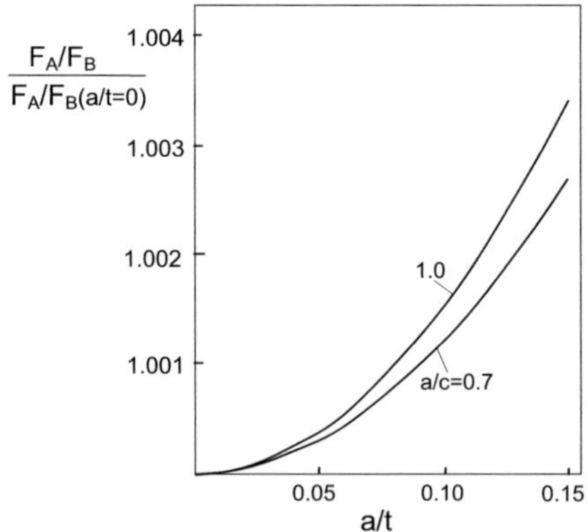

Fig. F3.4 Stress intensity factors for a semi-elliptical crack under remote tension; influence of the relative crack depth a/t on the ratio of the geometric functions.

F3.3.2 Semi-elliptical single cracks under bending load

Stress intensity factor solutions for cracks in bars under bending load are known in fracture mechanics literature. Also for this application let us use the equations proposed by Newman and Raju [F3.6]. The geometric functions F_A and F_B were determined via (F3.1.17) where now σ_0 is the outer fiber bending stress. For a wide aspect ratio range of $0.5 \leq a/c \leq 1$ and relative crack depths of $0 \leq a/t \leq 0.15$, they read

$$F_A = 1.107 - 0.5845\frac{a}{c} + 0.1436\left(\frac{a}{c}\right)^2 - 1.135\frac{a}{t} + 0.8454\left(\frac{a}{t}\right)^2$$
$$+ 0.311\frac{a}{c}\frac{a}{t} - 0.7621\left(\frac{a}{c}\right)\left(\frac{a}{t}\right)^2 \tag{F3.3.8}$$

$$F_B = 0.755 + 0.0232\frac{a}{c} - 0.0714\left(\frac{a}{c}\right)^2 - 0.569\frac{a}{t} + 0.739\left(\frac{a}{t}\right)^2$$
$$+ 0.0528\frac{a}{c}\frac{a}{t} - 0.541\left(\frac{a}{c}\right)\left(\frac{a}{t}\right)^2 \tag{F3.3.9}$$

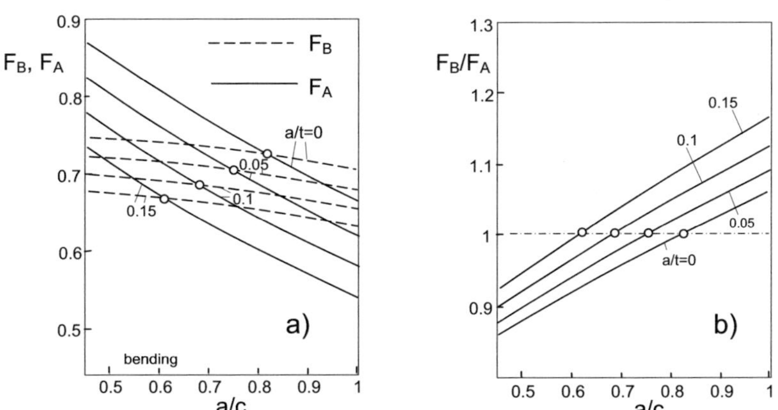

Fig. F3.5 Semi-elliptical surface crack under bending load; a) influence of the relative crack depth a/t and the aspect ratio a/c on the stress intensity factors, b) ratio of stress intensity factors, (circles indicate $F_A = F_B$).

The geometric functions and the ratio of F_A/F_B are plotted in Fig. F3.5 as a function of the relative crack depth a/t (t=thickness of the bar) and the aspect ratio a/c. Identical stress intensity factors at points A and B are indicated by the circles. For the aspect ratio $(a/c)_{eq}$ of the equilibrium ellipse it holds at small depths $a/t < 0.25$ approximately

$$\left(\frac{a}{c}\right)_{eq} \cong 0.812 - 1.382\frac{a}{t} + 0.1344\left(\frac{a}{t}\right)^2 \tag{F3.3.10}$$

For practical applications with a manageable number of terms, the geometric functions in the ranges of $0.5 \le a/c \le 1$ and $0 \le a/t \le 0.15$ may be simplified as

$$F_A \approx 1.02 - 0.36\frac{a}{c} - 0.866\frac{a}{t} \tag{F3.3.11}$$

$$F_B \approx 0.79 - 0.081\frac{a}{c} - 0.479\frac{a}{t} \tag{F3.3.12}$$

F3.4 A weight function for average stress intensity factors

In a similar way as outlined in [F3.7] for straight-trough cracks, a weight function description is possible also for semi-elliptical cracks under variable load distributed over the crack area. The average stress intensity factors may be written as

$$\overline{K}_{A,B} = \int\limits_{(S)} h_{A,B}\, dS \tag{F.3.4.1}$$

defining the weight functions h_A and h_B. Tables F3.1 and F3.2 compile data of these functions for stresses expressed as

$$\sigma = f(\varphi, \rho/R) \tag{F.3.4.2}$$

with the distance of the crack front from the origin

$$R = \sqrt{(a\sin\varphi)^2 + (c\cos\varphi)^2} \tag{F.3.4.3}$$

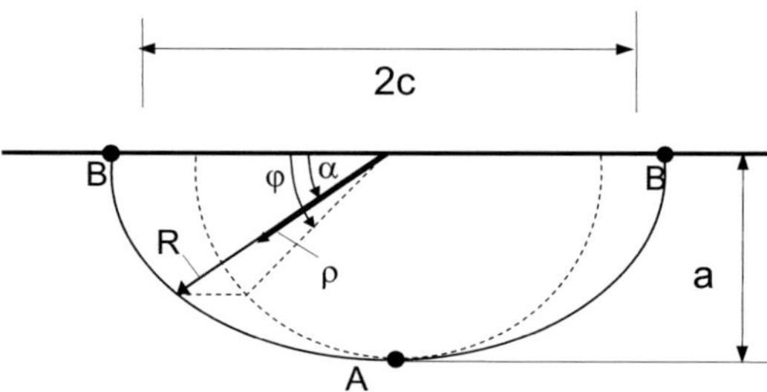

Fig. F3.6 Definition of the geometric data occurring in eqs.(F3.4.2)-(F3.4.5).

The parametric angle φ (angle with respect to the half-circle with radius a) is

$$\varphi = \arctan\left(\frac{c}{a}\tan\alpha\right) \tag{F.3.4.4}$$

and the area element dS

$$dS = \rho^2 d\alpha = a\,c\,d\varphi \qquad\qquad (F.3.4.5)$$

The geometric parameters are given in Fig. F3.6.

a/c		$\rho/R=0$	0.2	0.4	0.6	0.8	1.
0.8	$\alpha=0$	0.3782	0.3259	0.2602	0.1810	0.0928	0
	$\pi/8$	0.3782	0.3246	0.2753	0.2194	0.1594	0.09338
	$\pi/4$	0.3782	0.3284	0.3052	0.2894	0.2703	0.25734
	$3\pi/8$	0.3782	0.3493	0.3448	0.3551	0.3665	0.36949
	$\pi/2$	0.3782	0.3595	0.3678	0.3897	0.4075	0.40636
1.0	0	0.4036	0.3257	0.2601	0.1809	0.0927	0
	$\pi/8$	0.4036	0.3306	0.2795	0.2204	0.1526	0.07439
	$\pi/4$	0.4036	0.3417	0.3205	0.3041	0.2841	0.25398
	$3\pi/8$	0.4036	0.3586	0.3685	0.3927	0.4177	0.43356
	$\pi/2$	0.4036	0.3675	0.3910	0.4319	0.4748	0.50795
1.2	0	0.3574	0.3120	0.2491	0.1733	0.0888	0
	$\pi/8$	0.3574	0.3165	0.2671	0.2086	0.1414	0.05973
	$\pi/4$	0.3574	0.3221	0.3022	0.2867	0.2701	0.2358
	$3\pi/8$	0.3574	0.3354	0.3515	0.3872	0.4337	0.47867
	$\pi/2$	0.3574	0.3625	0.4026	0.4640	0.5334	0.60954

Table F3.1 Weight function for point A in normalized form $h_A(1-\rho/R)^{1/2}$.

a/c		$\rho/R=0$	0.2	0.4	0.6	0.8	1.
0.8	$\alpha=0$	0.3565	0.3346	0.3320	0.3524	0.3992	0.4543
	$\pi/8$	0.3565	0.3167	0.3330	0.3529	0.3629	0.34835
	$\pi/4$	0.3565	0.3132	0.2822	0.2496	0.2079	0.1647
	$3\pi/8$	0.3565	0.3073	0.2507	0.1848	0.1133	0.04057
	$\pi/2$	0.3565	0.3027	0.2383	0.1627	0.0814	0
1.0	0	0.5087	0.4567	0.4640	0.4900	0.5127	0.50795
	$\pi/8$	0.5087	0.4491	0.4386	0.4324	0.4289	0.43356
	$\pi/4$	0.5087	0.4372	0.3918	0.3445	0.2928	0.25398
	$3\pi/8$	0.5087	0.4211	0.3458	0.2587	0.1646	0.07439
	$\pi/2$	0.5087	0.4127	0.3246	0.2216	0.1110	0
1.2	0	0.6037	0.5668	0.5674	0.5832	0.5921	0.55643
	$\pi/8$	0.6037	0.5581	0.5377	0.5275	0.5168	0.50132
	$\pi/4$	0.6037	0.5435	0.4918	0.4380	0.3798	0.33956
	$3\pi/8$	0.6037	0.5280	0.4370	0.3322	0.2195	0.11826
	$\pi/2$	0.6037	0.5173	0.4064	0.2775	0.1391	0

Table F3.2 Weight function for point B in normalized form $h_B(1-\rho/R)^{1/2}$.

References F3

F3.1 Rice, J.R., Some remarks on elastic crack-tip stress fields, Int. J. Solids and Struct. **8**(1972), 751-758.

F3.2 Fett, T., Mattheck, C., Munz, D., Approximate weight functions for 2D and 3D problems, J. Engng. Analysis **6**(1989), 48-62.

F3.3 Fett, T., Munz, D., Stress intensity factors and weight functions, Computational Mechanics Publications, Southampton, 1997.

F3.4 Cruse, T.A., Besuner, P.M., Residual life prediction for surface cracks in complex structural details, J. of Aircraft **12**(1975), 369-375.

F3.5 Fett, T., Mattheck, C., Munz, D., On the accuracy of the Cruse-Besuner weight function approximation for semi-elliptical surface cracks, Int. J. Fract. **40**(1989), 307-313.

F3.6 Newman, J.C., Raju, I.S., An empirical stress intensity factor equation for the surface crack, Engng. Fract. Mech. **15**(1981), 185-192.

F3.7 Fett, T., Stress intensity factors, T-stresses, Weight functions, IKM 50, Universitätsverlag Karlsruhe, 2008.

F4

Surface cracks under residual stress loading

F4.1 Residual stress in a thin surface layer

Residual stresses caused by surface treatment of ceramics or generated in an ion exchange layer in glass are assumed here to be step-shaped. If b is the layer thickness and y the depth coordinate (Fig. F4.1), it holds

$$\sigma(y) = \begin{cases} \sigma_0 & for \quad y \le b \\ 0 & for \quad y > b \end{cases} \tag{F4.1.1}$$

with the surface stress σ_0. By using this stress distribution, stress intensity factors can be determined. For this purpose, it is assumed that the pre-existing surface cracks are semi-elliptically shaped (Fig. F4.1).

From the stresses of eq.(F4.1.1), stress intensity factors were computed according to the procedure described in Section F3. The related geometric functions $F_{A,B}$ are defined by

$$\overline{K}_{A,B} = \sigma_0 \, F_{A,B} \sqrt{\pi a} \; . \tag{F4.1.2}$$

In Figs. F4.2a and F4.2b, F_A and F_B are plotted versus b/a.

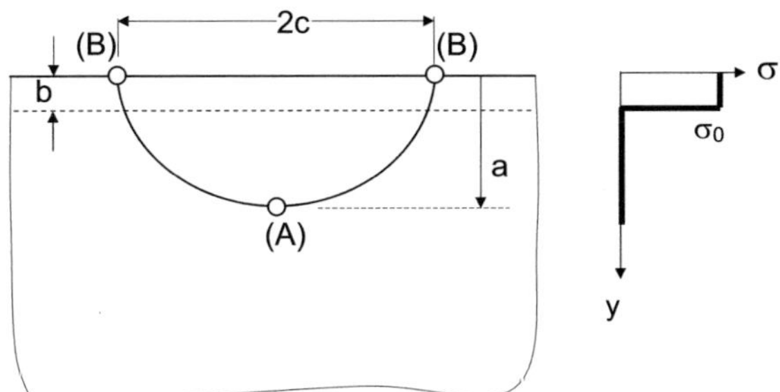

Fig. F4.1 Semi-elliptical surface cracks under a step-shaped residual stress distribution near the surface.

For small b/a-ratios shown in Fig. F4.2b, the straight-line relations can be concluded

$$F_A \cong \left(1.04 - 1.025\frac{a}{c} + 0.375\left(\frac{a}{c}\right)^2\right)\frac{b}{a} \tag{F4.1.3}$$

$$F_B \cong \left(2.37 - 0.725\frac{a}{c} + 0.125\left(\frac{a}{c}\right)^2\right)\frac{b}{a} \tag{F4.1.4}$$

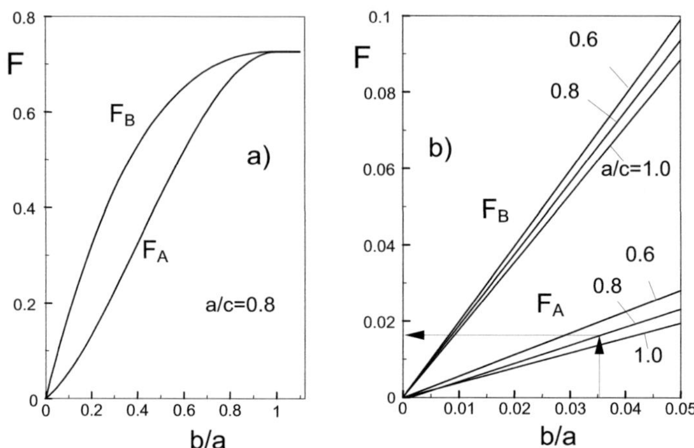

Fig. F4.2 Stress intensity factors for step-shaped loading a) geometric functions for *a/c*=0.8, b) geometric functions for small values of *b/a*.

F4.2 Concentrated crack-face loading

F4.2.1 Stress intensity factors

A crack is considered loaded in the centre region by a force *P*. This force can be caused by a constant pressure *p* distributed over an area near the crack centre that is small compared to the crack area. If this area is a circle of radius *b*, it holds

$$\overline{K}_{A,B} = \frac{2P}{(\pi a)^{3/2}} F_{A,B} = \frac{2P}{(\pi c)^{3/2}}\left(\frac{c}{a}\right)^{3/2} F_{A,B}$$

$$P = \tfrac{1}{2} p\pi b^2 \tag{F4.2.1}$$

The geometric functions F_A and F_B are given in Fig. F4.3 together with the ratio F_A/F_B. In the range of $0.7 \le a/c \le 1$, the geometric functions can be approximated by

$$F_A = 0.018 + 2.886\frac{a}{c} - 2.602\left(\frac{a}{c}\right)^2 + 0.732\left(\frac{a}{c}\right)^3 + 0.4063\frac{a}{c}\left(\frac{b}{a}\right)^2 + 0.016\left(\frac{a}{c}\frac{b}{a}\right)^2 \tag{F4.2.2}$$

$$F_B = -0.175 + 0.714\frac{a}{c} + 1.357\left(\frac{a}{c}\right)^2 - 0.565\left(\frac{a}{c}\right)^3 - 0.198\frac{a}{c}\left(\frac{b}{a}\right)^2 + 0.4334\left(\frac{a}{c}\frac{b}{a}\right)^2 \quad \text{(F4.2.3)}$$

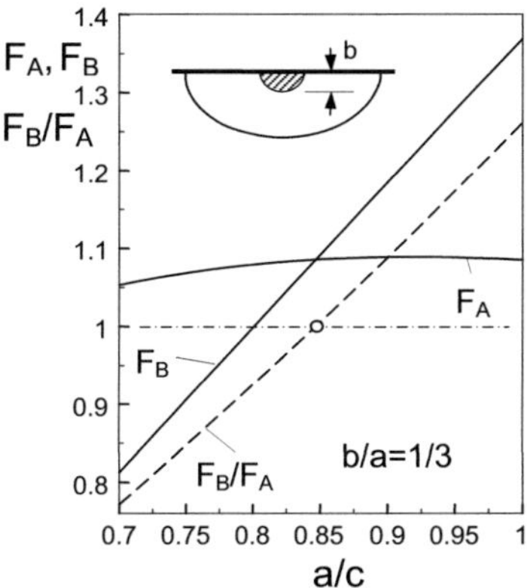

Fig. F4.3 Semi-elliptical surface crack loaded by a crack-face pressure p, distributed over a circle of radius $b=a/3$, geometric functions (solid curves) and ratio F_B/F_A (dashed curve).

For a value of $b/a=1/3$ relevant for Vickers indentation cracks, it holds simplified

$$F_A \cong 0.45 + 1.3764\frac{a}{c} - 0.7475\left(\frac{a}{c}\right)^2 \quad \text{(F4.2. 4)}$$

$$F_B \cong -0.52 + 1.9213\frac{a}{c} - 0.0362\left(\frac{a}{c}\right)^2 \quad \text{(F4.2.5)}$$

From (F4.2.4) and (F4.2.5) it becomes obvious that the condition $F_A=F_B$ is fulfilled for $(a/c)_{eq} = 0.847$. From this evaluation we have to expect a Vickers indentation crack after unloading as a semi-ellipse with an aspect ratio of 0.847.

There is an influence of b/a on the equilibrium aspect ratio, roughly approximated for $0 \le b/a \le 0.5$ by

$$(a/c)_{eq} \cong 0.836 + 0.12(b/a)^2 \quad \text{(F4.2.6)}$$

with the related geometric functions

$$F_A = F_B \cong 1.05 + 0.3315(b/a)^2 \quad \text{(F4.2.7)}$$

For the case of concentrated central force acting at the crack centre, $d=0$ (for d see Fig. F4.4a), the geometric functions for $0.8 \le a/c \le 1.2$ can be expressed by

$$F_A^{(P)} = 0.516 + 1.176\,a/c - 0.658(a/c)^2 \quad \text{(F4.2.8)}$$

$$F_B^{(P)} \cong -0.444 + 1.78\, a/c \qquad\qquad \text{(F4.2.9)}$$

F4.2.2 Stress intensity factors for point forces on the symmetry line

A point force P is considered with an offset d from the free surface (Fig. F4.4a). The geometric functions defined by eq.(F4.2.1) are plotted in Fig. F4.4b. The equilibrium aspect ratio is given in Fig. F4.4c. It may be fitted by the polynomial

$$(a/c)_{eq} = 0.837 + 0.267\, d/a - 0.721(d/a)^2 + 5.404(d/a)^3 \qquad \text{(F4.2.10)}$$

and the related stress intensity factors are

$$F_A = F_B = 1.045 + 0.141\, d/a + 0.547(d/a)^2 + 3.78(d/a)^3 - 0.032(d/a)^4 \qquad \text{(F4.2.11)}$$

From Fig. F4.4c it is obvious, that under a concentrated force a semi-circular crack will occur only for an offset of d/a=0.3.

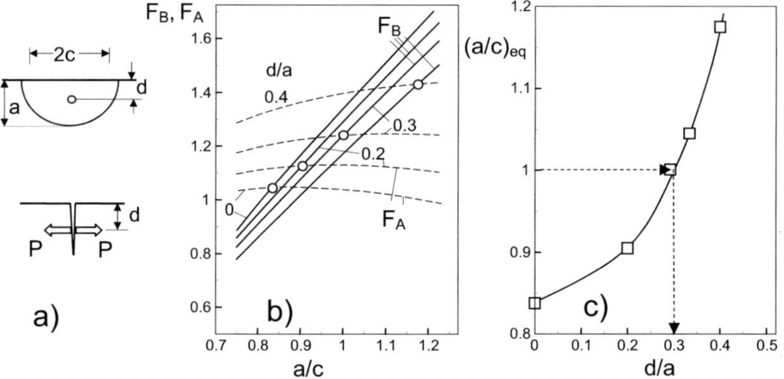

Fig. F4.4 a) Semi-elliptical crack loaded by a point force P exhibiting an offset d from the surface, b) geometric functions according to eq.(F4.2.1), c) equilibrium aspect ratio for F_A=F_B versus relative offset d/a.

F4.3 Loading by residual stresses constant over a semi-elliptic area

Knoop indentation tests carried out on brittle materials are accompanied by the generation of a half-elliptical surface crack below the indenter. For the fracture mechanics analysis of such cracks the stress intensity factor caused by the residual stress field is necessary. Sometimes the *local* stress intensity factor is used [F4.1] for this purpose. Unfortunately, there are hardly solutions available for a wide range of crack shapes and residual stress distributions.

In the following, average instead of local stress intensity factors will be used in order to compute the shape of the indentation crack after removal of the indentation load.

During a Knoop indentation test a residual stress zone develops below the contact area. According to the model proposed by Marshall [F4.2] a prolate spheroid was chosen for the shape of the irreversibly deformed 'plastic' zone with the ratio of the major axis b_1 and the depth b_2 (Fig. F4.5a). If σ_{res} is the residual stress assumed to be constant over the semi-elliptic cross section with the half-axes b_1 and b_2, the total force normal to the crack plane is

$$P_{res} = \tfrac{1}{2}\sigma_{res}\pi\, b_1 b_2 \tag{F4.3.1}$$

The stress intensity factors scaled with this force are plotted in Fig. F4.5b and F4.5c (solid curves) in the form

$$K_{A,B} = \frac{2P_{res}}{(\pi a)^{3/2}} F_{A,B} \tag{F4.3.2}$$

In order to apply the value of $b_1/b_2=3$ (widely used by Keer et al. [F4.1]), the geometric functions $F_{A,B}$ in (F4.3.2) were fitted for $0.7 \leq a/c \leq 1$ by

$$F_A \cong \sum_{i,j} C_{ij}\left(\frac{a}{c}\right)^i\left(\frac{b_1}{c}\right)^j \tag{F4.3.3}$$

with $C_{00}=0.5041$, $C_{10}=1.20$, $C_{20}=-0.6736$, $C_{01}=-0.2377$, $C_{02}=0.312$, $C_{11}=0.544$, $C_{12}=-0.885$, $C_{21}=-0.2683$, $C_{22}=0.429$ for (A) at $0.8<c/b_1<5$

$$F_B \cong -9.732 + 21.78\frac{c}{b_1} - 12.407\left(\frac{c}{b_1}\right)^2 + 2.76\frac{a}{c} - 0.37\left(\frac{a}{c}\right)^2 \quad \text{for } 0.8 \leq c/b_1 \leq 1 \tag{F4.3.4}$$

$$F_B \cong -0.4452 + 1.781\tfrac{a}{c} + \left(0.577 + 0.4395\tfrac{a}{c}\right)\exp[-2.149\sqrt{c/b_1 - 0.97}] \quad \text{for } c/b_1 > 1 \tag{F4.3.5}$$

From Fig. F4.5b it is clearly visible that the stress intensity factor at the deepest point of the semi-ellipse is smaller than the values at the surface in agreement with the analysis by Keer et al. [4.1]. The "equilibrium half-ellipse" for which the stress intensity factors at the deepest point (A) and the surface points (B) are identical must exhibit an aspect ratio of $(a/c)_{eq} \leq 0.76$ for $c/b_1 \leq 2$ as can be seen from Fig. F4.5d.

In order to show the influence of the shape of the spheroid, the axis ratio b_1/b_2 was varied. The result for $b_2=b_1/1.5$ is introduced in Fig. F4.5c by the dashed curves. These curves are slightly lower (about 4% for $a/c=1$ and $c/b_1>0.9$) than those for $b_2=b_1/3$. For a zone more concentrated near the surface, $b_2=b_1/6$, the results are represented by the dash-dotted curves. In the case of $c/b_1>1.05$ a rough representation is

$$F_B\left(\frac{b_1}{b_2}\right) \approx F_B(3) + 0.0533 - 0.16\frac{b_2}{b_1} \tag{F4.3.6}$$

Figure F4.6 shows those combinations of b_1/b_2 and c/b_1 which cause a certain equilibrium aspect ratio of the crack. From these results it can be concluded that an aspect ratio of $a/c=1$ (semi-circular crack) cannot be reached with constant residual stresses distributed over a semi-elliptical cross-section of $b_1/b_2>1$. An aspect ratio of $a/c=0.9$ is possible only for $b_1/b_2\leq1.1$ and $c/b_1<1.4$. Measurements by Marshall et al. [F4.3] showed an aspect ratio of $a/c\cong0.835$. This aspect ratio can be fulfilled in a wide range for c/b_1 by a value of $b_1/b_2=1.2$ indicating a nearly circular residual stress zone.

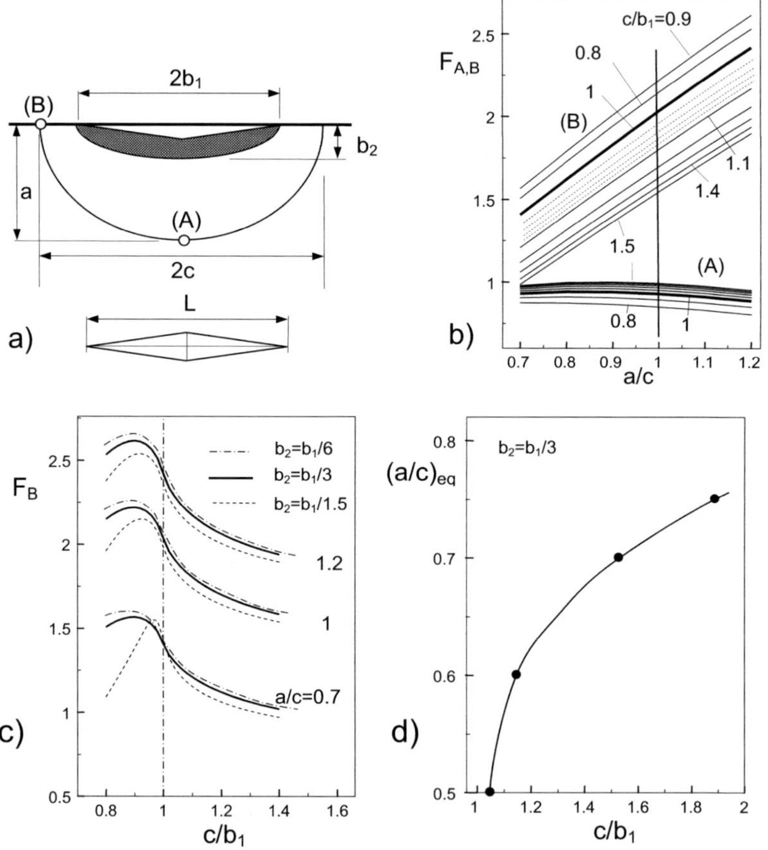

Fig. F4.5 a) Semi-elliptical crack loaded by a residual stress zone of length $2b_1$ and width $b_2=b_1/3$, b) stress intensity factors as functions of relative crack length c/b_1 and aspect ratio a/c, dashed lines between $c/b_1=1$ and 1.1 in steps of 0.02, c) stress intensity factor at the surface for differently chosen short half-axes of the residual zone, d) aspect ratio of the equilibrium semi-ellipse showing $F_A=F_B$.

From the existence of Knoop cracks with aspect ratios of $0.825 < a/c < 0.85$, it can be concluded that the ratio of b_1/b_2 should be clearly below $b_1/b_2 = 3$ because this value would result in $a/c < 0.8$ (Fig. F4.6a).

In this context it has to be mentioned that the semi-elliptical pressure distribution is not necessarily the same as the observed damaged zone. In the previous computations it means the effective area of <u>constant</u> pressure that yields the same stress intensity factors at the surface and the deepest point as the damaged zone with its varying pressure distribution would cause.

If residual surface stresses are also present in a specimen (Section F4.1), the Knoop indentation test may yield cracks with deviating aspect ratios.

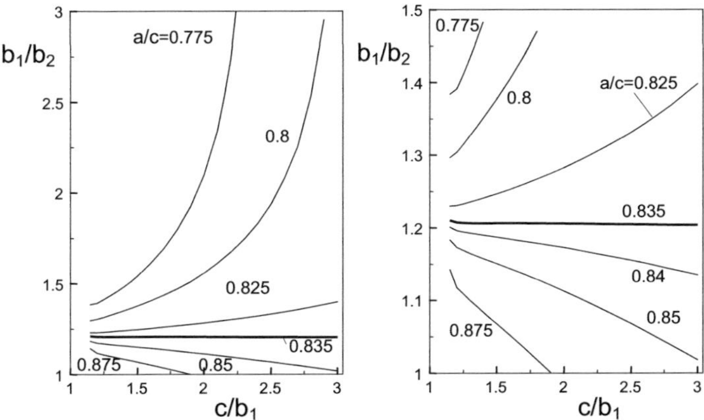

Fig. F4.6 Combinations of the aspect ratio of the residual stress zone, b_1/b_2, and relative crack width c/b_1, resulting in the same aspect ratio a/c of the crack.

The geometric functions for the special ratio of $b_1/b_2=1.2$ (Table F4.9) were fitted as

$$F_A \cong 0.583 + 1.124\frac{a}{c} - 0.665\left(\frac{a}{c}\right)^2 + \left(-1.054 + 1.734\frac{a}{c} - 0.724\left(\frac{a}{c}\right)^2\right)\frac{b_1}{c}$$
$$+ \left(2.894 - 4.659\frac{a}{c} + 2.038\left(\frac{a}{c}\right)^2\right)\left(\frac{b_1}{c}\right)^2 \tag{F4.3.7}$$

$$F_B \cong -0.448 + 1.79\tfrac{a}{c} + \left(0.5635 + 0.304\tfrac{a}{c}\right)\exp[-(3.024 - 0.5093\tfrac{a}{c})\sqrt{(c/b_1) - 0.97}] \tag{F4.3.8}$$

both for $0.7 \leq a/c \leq 1$ and $1 \leq c/b_1 \leq 5$. Figure F4.7 shows that the geometric functions for the case of $b_1/b_2=1.2$ with the circles indicating $(a/c)_{eq}$. Tables F4.1-F4.10 compile the results in a form which can easily be interpolated with respect of the parameters a/c,

c/b_1 and b_1/b_2. For this purpose it is recommended to interpolate the data with respect to $1/(c/b_1)$ instead of c/b_1.

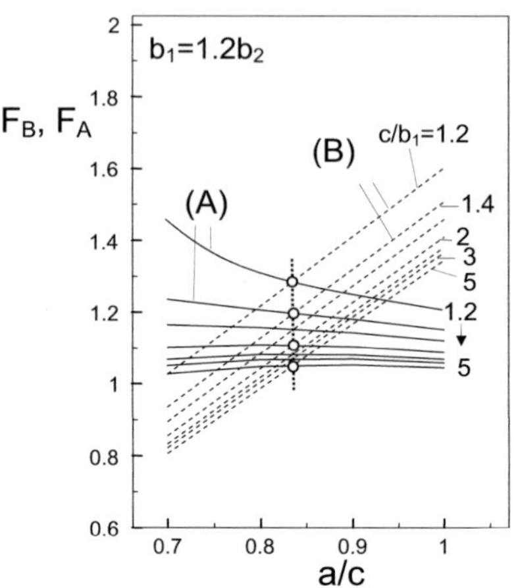

Fig. F4.7 Stress intensity factors for a semi-elliptical crack loaded by a residual stress zone of length $2b_1$ and width $b_2=b_1/1.2$; circles indicate the occurrence of $(a/c)_{eq}\cong0.835$ independent of the relative crack length c/b_1.

c/b_1	$a/c=0.7$	0.8	0.9	1
0.8	0.726	0.937	1.170	1.422
0.9	0.919	1.186	1.481	1.800
1.0	1.114	1.434	1.778	2.061
1.1	1.042	1.294	1.485	1.658
1.2	0.982	1.186	1.369	1.558
1.3	0.934	1.123	1.315	1.504
1.4	0.898	1.090	1.282	1.471
1.6	0.862	1.054	1.244	1.429
1.8	0.844	1.034	1.222	1.405
2.0	0.834	1.021	1.208	1.390
2.2	0.826	1.013	1.198	1.379
2.4	0.821	1.007	1.191	1.371
2.6	0.818	1.002	1.185	1.365
3.0	0.812	0.996	1.178	1.357
3.5	0.808	0.991	1.172	1.350
4	0.806	0.988	1.168	1.346
4.5	0.804	0.985	1.166	1.348
∞	0.802	0.980	1.158	1.336

Table F4.1 Normalized stress intensity factor F_B for $b_1=b_2$.

c/b_1	$a/c=0.7$	0.8	0.9	1	1.1	1.2	1.3
0.8	1.0893	1.406	1.714	1.958	2.174	2.376	2.565
0.9	1.379	1.704	1.929	2.138	2.339	2.532	2.719
1.0	1.436	1.596	1.787	1.981	2.172	2.358	2.538
1.1	1.151	1.347	1.544	1.737	1.924	2.105	2.280
1.2	1.059	1.255	1.449	1.639	1.823	2.002	2.173
1.3	1.005	1.199	1.391	1.579	1.761	1.937	2.107
1.4	0.968	1.161	1.351	1.537	1.717	1.892	2.061
1.5	0.942	1.133	1.322	1.506	1.685	1.859	2.027
1.8	0.893	1.081	1.266	1.448	1.625	1.796	1.962
2.0	0.874	1.060	1.244	1.425	1.600	1.771	1.936
2.2	0.860	1.045	1.228	1.408			
2.4	0.850	1.034	1.216	1.396			
2.6	0.842	1.025	1.207	1.386			
3.0	0.831	1.013	1.195	1.373			
3.5	0.822	1.004	1.184	1.362			
4.0	0.816	0.998	1.178	1.355			
4.5	0.812	0.993	1.173	1.351			
∞	0.802	0.980	1.158	1.336			

Table F4.2 Normalized stress intensity factor F_B for $b_1=1.5b_2$.

c/b_1	$a/c=0.7$	0.8	0.9	1
0.8	1.390	1.637	1.861	2.072
0.9	1.536	1.756	1.972	2.180
1.0	1.394	1.600	1.806	2.006
1.1	1.179	1.381	1.579	1.772
1.2	1.091	1.288	1.482	1.672
1.3	1.035	1.230	1.421	1.608
1.4	0.996	1.188	1.378	1.562
1.6	0.944	1.133	1.319	1.502
1.8	0.911	1.098	1.283	1.464
2.0	0.888	1.074	1.258	1.438
2.2	0.872	1.057	1.239	1.419
2.4	0.860	1.044	1.226	1.405
2.6	0.851	1.034	1.215	1.394
3.0	0.837	1.020	1.201	1.378
3.5	0.827	1.008	1.189	1.366
4.0	0.820	1.001	1.181	1.358
4.5	0.815	0.996	1.176	1.353
∞	0.802	0.980	1.158	1.336

Table F4.3 Normalized stress intensity factor F_B for $b_1=2b_2$.

c/b_1	$a/c=0.7$	0.8	0.9	1	1.1	1.2	1.3
0.8	1.506	1.728	1.941	2.146	2.342	2.531	2.711
0.9	1.565	1.789	2.007	2.217	2.419	2.614	2.801
1.0	1.407	1.621	1.830	2.033	2.229	2.418	2.601
1.1	1.207	1.410	1.609	1.802	1.989	2.170	2.344
1.2	1.118	1.315	1.509	1.698	1.882	2.059	2.231
1.3	1.059	1.254	1.445	1.631	1.812	1.988	2.158
1.4	1.017	1.209	1.398	1.583	1.763	1.937	2.105
1.6	0.985	1.176	1.363	1.546	1.725	1.898	2.065
1.8	0.924	1.111	1.296	1.476	1.653	1.824	1.990
2.0	0.899	1.084	1.268	1.448	1.623	1.794	1.959
2.2	0.881	1.065	1.248	1.427			
2.4	0.867	1.051	1.233	1.412			
2.6	0.857	1.040	1.221	1.399			
3.0	0.842	1.024	1.205	1.383			
3.5	0.830	1.011	1.192	1.369			
4.0	0.822	1.003	1.183	1.361			
4.5	0.817	0.998	1.178	1.355			
∞	0.802	0.980	1.158	1.336			

Table F4.4 Normalized stress intensity factor F_B for $b_1=3b_2$.

c/b_1	$a/c=0.7$	0.8	0.9	1
0.8	1.539	1.759	1.971	2.175
0.9	1.579	1.805	2.023	2.233
1.0	1.416	1.632	1.842	2.046
1.1	1.219	1.422	1.621	1.815
1.2	1.128	1.326	1.520	1.710
1.3	1.069	1.263	1.454	1.641
1.4	1.026	1.217	1.407	1.591
1.6	0.967	1.156	1.342	1.525
1.8	0.929	1.116	1.301	1.481
2.0	0.903	1.088	1.272	1.452
2.2	0.884	1.068	1.251	1.430
2.4	0.870	1.053	1.235	1.414
2.6	0.859	1.042	1.223	1.402
3.0	0.843	1.025	1.206	1.384
3.5	0.831	1.012	1.193	1.370
4.0	0.823	1.004	1.184	1.361
4.5	0.817	0.998	1.178	1.355
∞	0.802	0.980	1.158	1.336

Table F4.5 Normalized stress intensity factor F_B for $b_1=4b_2$.

c/b_1	$a/c=0.7$	0.8	0.9	1
0.8	0.757	0.941	1.136	1.339
0.9	0.958	1.191	1.437	1.695
1.0	1.182	1.470	1.772	1.928
1.1	1.388	1.690	1.791	1.487
1.2	1.536	1.734	1.469	1.366
1.3	1.612	1.474	1.362	1.297
1.4	1.562	1.361	1.298	1.250
1.6	1.295	1.254	1.223	1.191
1.8	1.213	1.198	1.179	1.154
2.0	1.166	1.163	1.150	1.130
2.2	1.135	1.138	1.130	1.112
2.4	1.114	1.121	1.115	1.100
2.6	1.098	1.107	1.104	1.090
3.0	1.076	1.089	1.088	1.076
3.5	1.059	1.075	1.076	1.065
4	1.048	1.066	1.068	1.059
4.5	1.041	1.060	1.063	1.054
∞	1.017	1.036	1.041	1.034

Table F4.6 Normalized stress intensity factor F_A for $b_1=b_2$.

c/b_1	$a/c=0.7$	0.8	0.9	1
0.8	1.137	1.075	1.033	0.996
0.9	1.088	1.056	1.030	1.002
1.0	1.056	1.045	1.030	1.010
1.1	1.041	1.041	1.032	1.015
1.2	1.034	1.039	1.033	1.018
1.3	1.029	1.038	1.034	1.021
1.4	1.026	1.037	1.035	1.023
1.6	1.022	1.036	1.036	1.026
1.8	1.020	1.036	1.038	1.029
2.0	1.018	1.036	1.039	1.030
2.2	1.018	1.036	1.039	1.031
2.4	1.017	1.036	1.040	1.032
2.6	1.017	1.036	1.040	1.032
3.0	1.016	1.037	1.041	1.033
3.5	1.016	1.037	1.042	1.034
4	1.015	1.037	1.042	1.035
4.5	1.015	1.037	1.043	1.035
∞	1.017	1.036	1.041	1.034

Table F4.7 Normalized stress intensity factor F_A for $b_1=2b_2$.

c/b_1	$a/c=0.7$	0.8	0.9	1
0.8	0.874	0.873	0.864	0.849
0.9	0.904	0.909	0.905	0.892
1.0	0.930	0.940	0.939	0.928
1.1	0.947	0.959	0.960	0.950
1.2	0.958	0.973	0.974	0.965
1.3	0.967	0.983	0.985	0.976
1.4	0.973	0.991	0.994	0.985
1.6	0.983	1.002	1.006	0.998
1.8	0.990	1.010	1.014	1.006
2.0	0.994	1.015	1.020	1.012
2.2	0.998	1.019	1.024	1.017
2.4	1.001	1.022	1.027	1.020
2.6	1.003	1.024	1.030	1.022
3.0	1.006	1.028	1.033	1.026
3.5	1.008	1.030	1.036	1.029
4	1.010	1.032	1.038	1.031
4.5	1.011	1.034	1.040	1.032
∞	1.017	1.036	1.041	1.034

Table F4.8 Normalized stress intensity factor F_A for $b_1=3b_2$.

c/b_1	$a/c=0.7$	0.8	0.9	1
0.8	0.805	0.814	0.812	0.802
0.9	0.853	0.865	0.865	0.856
1.0	0.892	0.906	0.908	0.900
1.1	0.916	0.933	0.936	0.928
1.2	0.933	0.951	0.955	0.947
1.3	0.946	0.965	0.969	0.961
1.4	0.956	0.975	0.980	0.972
1.6	0.970	0.990	0.995	0.988
1.8	0.980	1.000	1.006	0.999
2.0	0.987	1.008	1.013	1.006
2.2	0.992	1.013	1.019	1.012
2.4	0.995	1.017	1.023	1.016
2.6	0.998	1.020	1.026	1.019
3.0	1.003	1.025	1.031	1.024
3.5	1.006	1.029	1.035	1.028
4.0	1.009	1.031	1.037	1.031
4.5	1.010	1.033	1.040	1.032
∞	1.017	1.036	1.041	1.034

Table F4.9 Normalized stress intensity factor F_A for $b_1=4b_2$.

c/b_1	a/c=0.7	0.8	0.9	1	0.7	0.8	0.9	1
	F_A				F_B			
1.2	1.457	1.308	1.251	1.206	1.027	1.217	1.411	1.601
1.3	1.302	1.246	1.209	1.174	0.969	1.163	1.356	1.544
1.4	1.235	1.206	1.180	1.151	0.935	1.128	1.319	1.506
1.6	1.165	1.157	1.143	1.121	0.894	1.085	1.274	1.458
1.8	1.126	1.128	1.119	1.101	0.871	1.059	1.246	1.428
2.0	1.103	1.109	1.103	1.088	0.856	1.042	1.227	1.409
2.5	1.068	1.081	1.080	1.068	0.834	1.018	1.201	1.380
3.0	1.051	1.067	1.069	1.058	0.822	1.005	1.187	1.365
3.5	1.041	1.059	1.062	1.052	0.816	0.998	1.179	1.357
4.0	1.035	1.054	1.058	1.049	0.812	0.993	1.174	1.351
5.0	1.027	1.048	1.053	1.044	0.806	0.987	1.168	1.345
∞	1.017	1.036	1.041	1.034	0.802	0.980	1.158	1.336

Table F4.10 Normalized stress intensity factor F_B for b_1=1.2b_2.

F4.4 Cruciform cracks under remote tension

The semi-circular cruciform crack is illustrated by the inset of Fig. F4.8. The stress intensity factors K_{cr} (subscript "cr" for "cruciform") under remote uniaxial tension σ_0 were reported for instance by Keer et al. [F4.1] and Murakami and Sakae [F4.4]. The solution from [F4.4] is plotted in Fig. F4.8 by the circles with the shape function Y_{cr} for the local stress intensity factors defined by

$$K_{cr} = \sigma_0 \sqrt{\pi a} Y_{cr},$$
(F4.4.1)

with the angle φ in radian.

For $\varphi \leq \pi/2$ (again with φ=0 at the surface), the numerical results may be fitted by

$$Y_{cr} = 0.769 - 0.233\varphi + 0.182\varphi^2 - 0.0391\varphi^3$$
(F4.4.2)

For $\varphi > \pi/2$ it trivially holds

$$Y_{cr}(\pi - \varphi) = Y_{cr}(\varphi)$$
(F4.4.3)

The fit relation is plotted in Fig. F4.8 as the solid curve. Since at $\varphi = 90°$ the two cracks intersect, the stress singularity must deviate from the power ½ at this location and a stress intensity factor description is no longer possible directly at $\varphi = 90°$ (see Section F2.2). The geometric function for a single semi-circular crack according to (F4.4.1) is shown in Fig. F4.8 by the dashed curve.

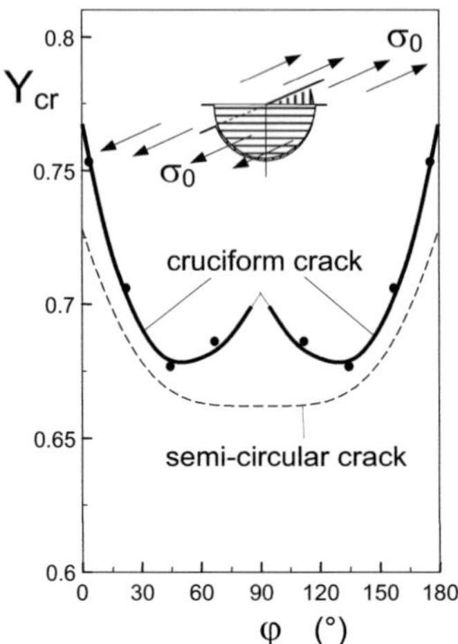

Fig. **F4.8** Comparison of local stress intensity factors for single semi-circular and cruciform semi-circular surface cracks loaded by uniaxial tension.

The ratio of the geometric functions for the two averaged stress intensity factors of the single and the cruciform semi-circular cracks according to (F3.1.17) is

$$\frac{F_{cr,A,B}}{F_{A,B}} \cong 1.03 \tag{F4.4.4}$$

i.e. an influence of the cruciform cracks is in terms of averaged stress intensity factors rather small.

As the consequence of roughly the same stress intensity factor increase for point A and point B, the aspect ratios of the cruciform cracks should be the same as for single semi-elliptical cracks.

References F4

F4.1 Keer, L.M., Farris, T.N., Lee, J.C., Knoop and Vickers indentation in ceramics analyzed as a three-dimensional fracture, J. Am. Ceram. Soc. **69**(1986), 392-96.

F4.2 Marshall, D.B., Controlled flaws in ceramics: A comparison of Knoop and Vickers indentation, J. Am. Ceram. Soc. **66**(1983), 127-131.

F4.3 Marshall, D.B., Evans, A.G., Yakub, B.T.K., Tien, J.W., Kino, G.S., The nature of machining damage in brittle materials, Proc. R. Soc. Lond. A**385**(1983), 461-75.

F4.4 Murakami, Y., Sakae, C., Analysis of stress intensity factors for three-dimensional cruciform cracks, Int. J Fract. **66**(1994), 339-355.

F5

Rectangular surface cracks

F5.1 Stress intensity factors

A rectangular surface crack of depth a and width $2c$ is illustrated in Fig. F5.1a. Such cracks may be introduced in test specimens by the focussed ion beam (FIB) technique resulting in a rectangular or a slightly trapezoidal cross section.

Under tensile loading, the stress intensity factors at points (A) and (B) are defined by

$$K_{A,B,tension} = \sigma F_{A,B} \sqrt{\pi a} \qquad (F5.1.1)$$

with the geometric functions $F_{A,B}$.

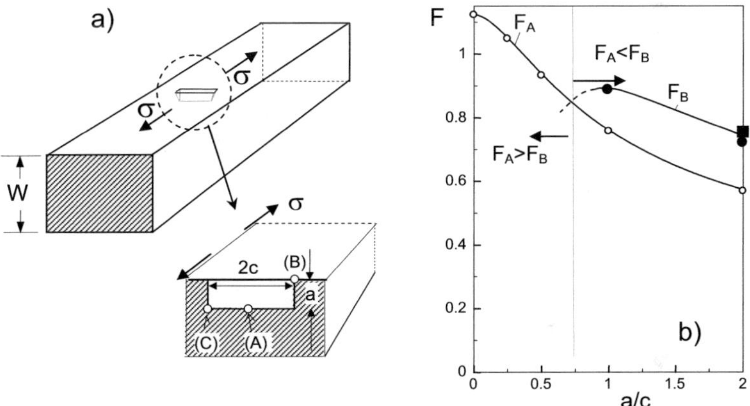

Fig. F5.1 a) Rectangular crack under tension, b) geometric functions at point (B) and point (A); open circles: geometric functions for the stress intensity factor at point (A) by Isida et al.[F5.1], solid circles: stress intensity factors at point (B) by Noguchi and Smith [F5.2], square: by Murakami [F5.3].

On the basis of tabulated data (open circles in Fig. F5.1b) taken from tables reported by Isida et al. [F5.1] a fit relation is derived

$$F_A = \frac{F_0}{\left[1 + \dfrac{25}{8}\left(\dfrac{a}{c}\right)^{7/4}\right]^{0.276}} \qquad (F5.1.2)$$

as introduced in Fig. F5.1b as the solid line. The value $F_0 = 1.1215$ is the well-known solution for the edge-cracked half-space. The solid circles in Fig. F5.1b are numerical data by Noguchi and Smith [F5.2] for the geometric function F_B. The square represents a single result by Murakami [F5.3]. These F_B-values together with the trivial solution of $F_B=0$ for $a/c=0$ (edge crack) show that the stress intensity factor at point (A) is the maximum value only for roughly $a/c<3/4$. From this point it is recommended to produce notches with a/c as small as possible.

For crack depths very small compared to the specimens thickness, $a<<W$, the tensile solutions are also applicable for bending. For relative notch depths of $0<a/W<0.1$, it is proposed to apply the correction factors H_A and H_B derived by Newman and Raju [F5.4] for semi-elliptical surface cracks with

$$H_A = 1-[1.22+0.12(a/c)]\,a/W \qquad\qquad\text{(F5.1.3)}$$

and

$$H_B = 1-[0.34+0.11(a/c)]\,a/W \qquad\qquad\text{(F5.1.4)}$$

Then it holds

$$K_{A,B,bending} = \sigma F_{A,B} H_{A,B} \sqrt{\pi a} \qquad\qquad\text{(F5.1.5)}$$

Tolerating maximum deviations of 1.5%, a simple relation is suggested for $a/c\leq0.5$ by

$$F_A \cong F_0 - 0.36\frac{a}{c} \qquad\qquad\text{(F5.1.6)}$$

Small deviations from the rectangular shape may be caused during the FIB-cutting procedure. After notch preparation a slightly trapezoidal cross-section has to be expected as shown in Fig. F5.2a.

The relations for the rectangular crack can also be applied for slightly trapezoidal cracks. The maximum stress intensity factor of a rectangular crack of width $2c$ must be smaller than the trapezoidal crack and the rectangular crack with a width of $2c_1$ must exceed it, i.e.

$$F_A(a/c) < F_{A,trapez} < F_A(a/c_1) \qquad\qquad\text{(F5.1.7)}$$

An approximation for the geometric function of the trapezoidal crack is given by

$$F_{A,trapez} \cong \tfrac{1}{2}[F_A(a/c)+F(a/c_1)] \qquad\qquad\text{(F5.1.8)}$$

For an angle of $\beta=10°$, the geometric function according to (F5.1.8) as well as the two limit cases expressed by (F5.1.7) are plotted in Fig. F5.2b. Up to $a/c=3/4$, the maximum possible error of (F5.1.8) is less than 2%.

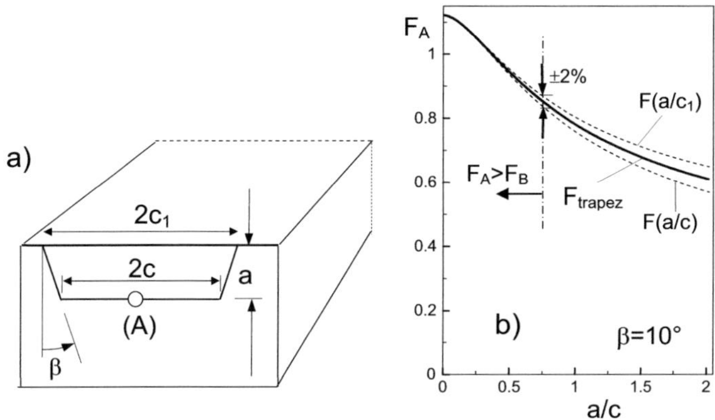

Fig. F5.2 a) A bar with a crack of slightly trapezoidal cross section, b) geometric function according to eq.(F5.1.8) with limit cases (dashed curves) given by eq.(F5.1.7).

F5.2 Compliance

Similar to eq.(E4.3.1), the compliance caused by the rectangular surface crack (Fig. F5.3) is

$$C_{rect} = \frac{4c}{E'} \int_0^a \frac{1}{P^2} K^2(a')\,da' \tag{F5.2.1}$$

For the special case of $a/c \ll 1$, the stress intensity factor originally derived for point (A), can be assumed to be constant over the crack width $2c$. Under this assumption, introducing of eqs.(F5.1.1) and (F5.1.2) yields for a 4-point bending test with outer loading point span S_2 and inner span S_1 for $a \ll W$

$$C_{rect} = \frac{9\pi F_0^2 a^2 c (S_2 - S_1)^2}{2B^2 W^4 E'} \, {}_2F_1\left(\tfrac{8}{7}, 0.552; \tfrac{15}{7}, -\tfrac{25}{8}\left(\tfrac{a}{c}\right)^{7/4}\right) \tag{F5.2.2}$$

where ${}_2F_1(\ )$ is the hypergeometric function.
A series expansion of (F5.2.2) reads

$$C_{rect} = \frac{9\pi F_0^2 a^2 c (S_2 - S_1)^2}{2B^2 W^4 E'} + O\left(\left(\tfrac{a}{c}\right)^3\right) \tag{F5.2.3}$$

which for the small *edge crack* ($a/c=0$, $c=B$, $a/W \to 0$) results in the well-known relation

113

$$C_{edge} = \frac{9\pi F_0^2 a^2 (S_2 - S_1)^2}{2BW^4 E'} \tag{F5.2.4}$$

Conseuently, it can be written

$$C_{rect} \cong C_{edge} \frac{c}{B} \tag{F5.2.5}$$

where the approximation sign accounts for the fact that C_{rect} is an upper limit solution since the maximum K along the crack front was used. The equality sign can be used at least for $a/c<0.1$.

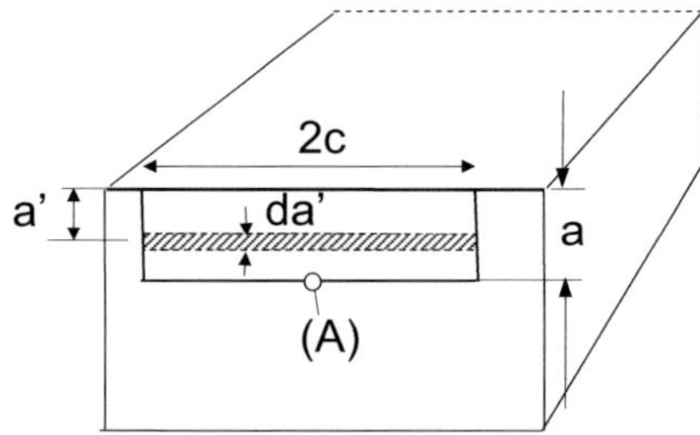

Fig. F5.3 Rectangular surface crack with the definition of a virtual crack increment da'.

References F5

F5.1 Isida, M. Yoshida, T., Noguchi, H., A rectangular crack in an infinite solid, a semi-infinite solid and a finite-thickness plate subjected to tension, Int. J. Fract. **52**(1991), 79-90.

F5.2 Noguchi, H., Smith, R., An analysis of a semi-infinite solid with three-dimensional cracks, Engng. Fract. Mech. **52**(1995), 1-14.

F5.3 Murakami, Y., Analysis of stress intensity factors of mode I, II, III of inclined surface crack of arbitrary shape, Engng. Fract. Mech. **22**(1985), 101-114.

F5.4 Newman, J.C., Raju, I.S., An empirical stress intensity factor equation for the surface crack, Engng. Fract. Mech. **15**(1981), 185-192.

F6

Semi-elliptical cracks ahead of notches

In this section the slender notch of length a_0 is addressed once more (see Section C19 in [F6.1]). For an edge crack emanating from the notch root a simple relation for the stress intensity factor K was derived

$$K / K^* = \tanh[2.243 A \sqrt{a/R}] \qquad \text{(F6.1)}$$

where R is the notch root radius and a the depth of the edge crack emanating from the notch. In this relation K^* is the stress intensity factor of an edge crack with the total length $a_{\text{total}} = a_0 + a$. This relation can be applied for the evaluation of fracture toughness of ceramic materials with the natural defects at the notch root modelled by an effective straight through-the-thickness crack. For special applications, crack-like defects in ceramics may be better described by semi-circular or semi-elliptical cracks.

For not too large values of a/R, the crack problem can be simplified significantly. If we restrict our considerations to $a/R<1$, we can ignore the curvature of the free surface at the notch root and compute the stress intensity factor by loading a crack in a flat plate with the same stress distribution as occurring in front of the notch (Fig. F6.1).

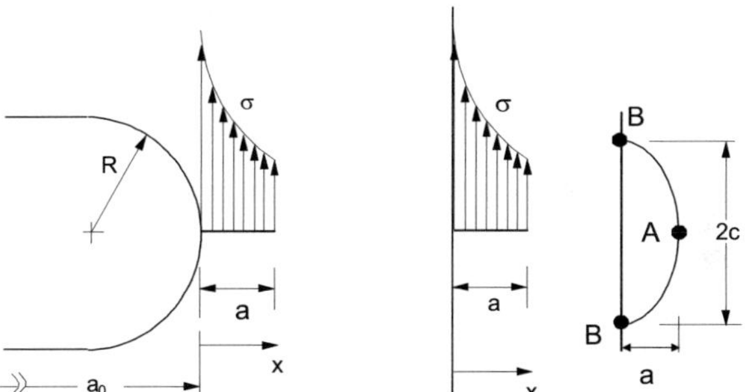

Fig. F6.1 Approximation of the notch problem by loading the same crack in a flat plate with the same stress profile.

The geometric functions for the deepest point (A) and the surface points (B) may be defined by

$$K_{A,B} = \sigma_{\max} Y_{A,B} \sqrt{a} \qquad \text{(F6.2)}$$

where $\sigma_{\max}$ is the maximum normal stress at the notch root (for details see [F6.2]).

The geometric functions for the local stress intensity factors at the deepest point (A) and the surface points (B) of the semi-ellipse [F6.2] are shown in Fig. F6.2.

Cracks with combinations of a/c and a/R resulting in a ratio $Y_B/Y_A < 1$ will fail under increasing load without stable crack extension in c-direction. In case of combinations located above the line $Y_B/Y_A = 1$, cracks will first extend in c-direction when $K_B=K_{Ic}$ is reached until the stress intensity factor at point A equals the stress intensity factor at point B. Then, spontaneous failure follows. The critical aspect ratios $(a/c)_c$ at failure can be obtained from the intersection of the curves in Fig. F6.3a with the value $Y_B/Y_A = 1$ (dash-dotted line). Figure F6.3b shows the critical aspect ratio as a function of the relative crack depth a/R.

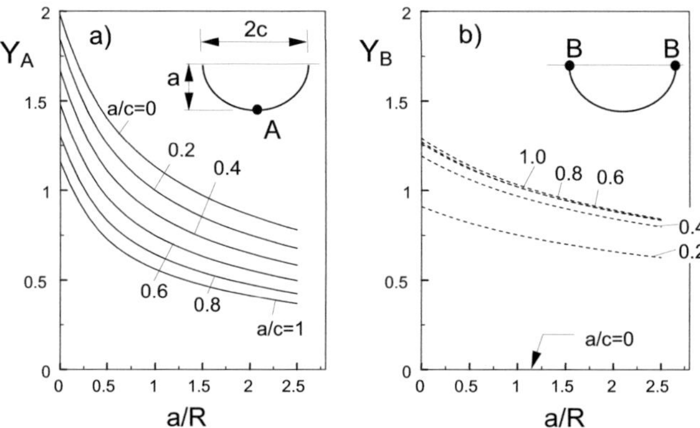

Fig. F6.2 Geometric functions for stress intensity factors of semi-elliptical surface cracks under a notch stress distribution.

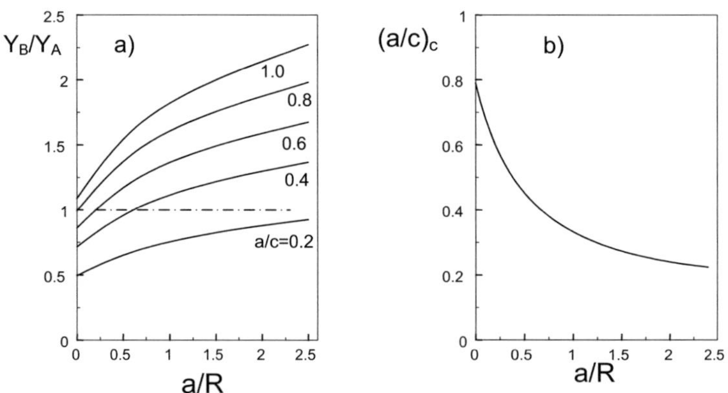

Fig. F6.3 a) Ratio of geometric functions, b) critical value for the aspect ratio a/c.

116

The critical value of the geometric function, Y_c, normalized on the edge-crack solution ($a/c = 0$) can be described by

$$Y_c / Y_{edge} = g(a/R) \cong \frac{2}{3} + 0.178\left[1 - \exp\left(-1.64\frac{a}{R}\right)\right] \qquad (F6.3)$$

Finally, a relation similar to eq.(F6.1) is proposed, which reads

$$K / K^* \cong \tanh[2.243 g(a/R)\sqrt{a/R}] \qquad (F6.4)$$

with the function $g(a/R)$ defined by eq.(F6.3). This relation is plotted in Fig. F6.4 as the solid line together with the relation for an edge crack at the notch root (dashed curve). Both of the curves show the same characteristic shape. Therefore, the curve for the semi-elliptical crack can be approximated by a straight-through crack with an effective length a_{eff} by

$$K_{semi-ell}(a) = K_{edge}(a_{eff}) \qquad (F6.5)$$

with

$$a_{eff} \cong \tfrac{5}{7} a \qquad (F6.6)$$

In the case of an array of semi-elliptical cracks, the factor of about 5/7 will increase against 1.

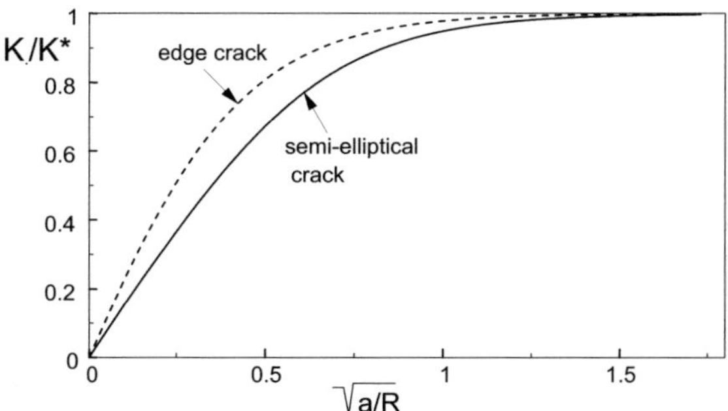

Fig. F6.4 Representation of eqs.(F6.1) and (F6.4).

References F6

F6.1 Fett, T., Stress intensity factors, T-stresses, Weight functions, IKM 50, Universitätsverlag Karlsruhe, 2008.
F6.2 Fett, T., Estimated stress intensity factors for semi-elliptical cracks in front of narrow circular notches, Engng. Fract. Mech. **64**(1999), 357-62.

F7

Miscellaneous problems on curved cracks

F7.1 Ring crack in a non-radialsymmetric stress field

F7.1.1 Linear stress distribution

In Section D2.1 the weight functions for the ring-shaped crack were derived for the case of rotational symmetry, i.e. they do not depend on the angle along the circumference. In this section, it is checked whether these weight functions are applicable also to non-radialsymmetric stresses.

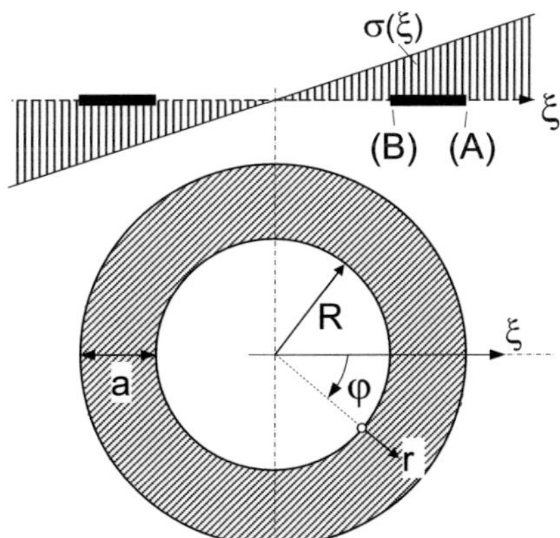

Fig. F7.1 Ring-shaped crack under a linear sress distribution.

As an example, the linear stress distribution

$$\sigma = \sigma_1 \frac{\xi}{R+a} \qquad \text{(F7.1.1)}$$

may be considered with the stress value $\sigma=\sigma_1$ occurring at the outer radius (i.e. at $\xi=R+a$). The stress along the radial r-axis with origin at the inner circumference of the ring (see Fig. F7.1) is then given by

$$\sigma = \sigma_1 \frac{R+r}{R+a} \cos\varphi \qquad \text{(F7.1.2)}$$

Introducing this stress in the weight function integral

119

$$K_{(A)} = \int_0^a h_{(A)}\sigma(r)dr \, , \quad K_{(B)} = \int_0^a h_{(B)}\sigma(r)dr \qquad \text{(F7.1.3)}$$

and use of the weight function (D2.1.3)

$$h_A\sqrt{a} = \frac{2}{\sqrt{\pi}\,\sqrt{1-\frac{r}{a}}\,\sqrt{\frac{a}{R}+1}}\left(\frac{1+\frac{r}{R}}{\sqrt{2+\frac{a}{R}+\frac{r}{R}}} - \frac{1-\sqrt{\frac{r}{a}}}{\sqrt{\frac{a}{R}+2}}\right) \qquad \text{(F7.1.4)}$$

result in the solid curve plotted in Fig. F7.2a.

The dotted line represents a solution for the linear stress distribution proposed by Tada et al. [F7.1]. For the outer crack tip, it reads

$$K_A = \sigma_1\sqrt{\frac{\pi a}{2}}\cos\varphi\left[1 - 0.401\frac{a}{R+a} - 0.065\left(\frac{a}{R+a}\right)^2 + 0.066\left(\frac{a}{R+a}\right)^3\right] \qquad \text{(F7.1.5)}$$

The ratio of the solution (F7.1.4) via approximate weight function and the literature solution (F7.1.5) is shown in Fig. F7.2b. For $a/R \leq 1$, the error of the weight function evaluation is less than 5%. Use of the radial-symmetric weight function should not be extended to larger a/R.

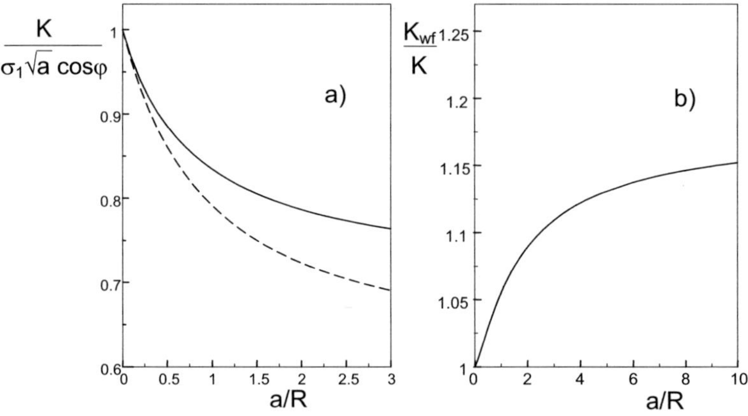

Fig. F7.2 a) Stress intensity factors for a linear stress distribution according to (F7.1.1, F7.1.1), solid curve: computed with (F7.1.4), dashed curve: solution (F7.1.3) given by Tada [F7.1].

The derivation of an improved weight function is possible by adjusting h to the reference stress intensity factor solution (F7.1.5).

If the weight functions are approximated by

$$h_{(A)} = \sqrt{\frac{2}{\pi a}}\left(\sqrt{\frac{\frac{r}{a}}{1-\frac{r}{a}}} + D_{(A)}\sqrt{\frac{r}{a}\left(1-\frac{r}{a}\right)}\right) \qquad (F7.1.6)$$

$$h_{(B)} = \sqrt{\frac{2}{\pi a}}\left(\sqrt{\frac{1-\frac{r}{a}}{\frac{r}{a}}} + D_{(B)}\sqrt{\frac{r}{a}\left(1-\frac{r}{a}\right)}\right) \qquad (F7.1.7)$$

it results

$$D_{(A)} = -\frac{1.604\alpha + 3.468\alpha^2 + 1.6\alpha^3}{(1+\alpha)^3} \qquad (F7.1.8)$$

$$D_{(B)} = 4\left[\frac{1}{\sqrt{1+\alpha}}\left(0.573 + \frac{0.427}{(1+\alpha)^{1/4}} - 0.26\frac{\alpha^{5/2}}{(1+\alpha)^4}\right) - 1\right] \qquad (F7.1.9)$$

F7.1.2 Arbitrary stress distribution

A Taylor series expansion of the stresses along the ξ-axis reads

$$\sigma(\xi) = \sigma(0) + \frac{d\sigma}{d\xi}\bigg|_{\xi=0}\xi + \tfrac{1}{2}\frac{d^2\sigma}{d\xi^2}\bigg|_{\xi=0}\xi^2 + \dots \qquad (F7.1.10)$$

For the first two terms, handbook solutions exist. The constant term is addressed in detail in Section D2.1. The linear term can be evaluated by use of (F7.1.5) according to Tada et al. [F7.1]. Only the higher-order terms summarised by the remaining stress distribution $\Delta\sigma(\xi)$ of

$$\Delta\sigma(\xi) = \sigma(\xi) - \sigma_0 - \sigma_1\frac{\xi}{a+R} \qquad (F7.1.11)$$

need to be evaluated by the non-radialsymmetric weight functions (F7.1.6) and (F7.1.7) with coefficients (F7.1.8) and (F7.1.9).

As an example of application, let us consider a ring crack located in the plane of a straight crack (Fig. F7.3).

If the ring crack is located in the prospective crack plane at a distance d from the tip of the straight (main) crack, the normal stress on the ring crack is given by

$$\sigma_n(\xi) = \frac{K_{appl}}{\sqrt{2\pi(d+\xi)}} \qquad (F7.1.12)$$

In (F7.1.12) K_{appl} is the stress intensity factor of the straight crack. It is assumed here that the stress intensity factor of the main crack is not influenced significantly by the existence of the ring crack. From a Taylor series expansion with respect to ξ, the first stress terms result as

$$\sigma_0 = \frac{K_{appl}}{\sqrt{2\pi d}} \tag{F7.1.13}$$

and

$$\sigma_1 = -\frac{K_{appl}}{\sqrt{2\pi d}}\frac{R+a}{2d} \tag{F7.1.14}$$

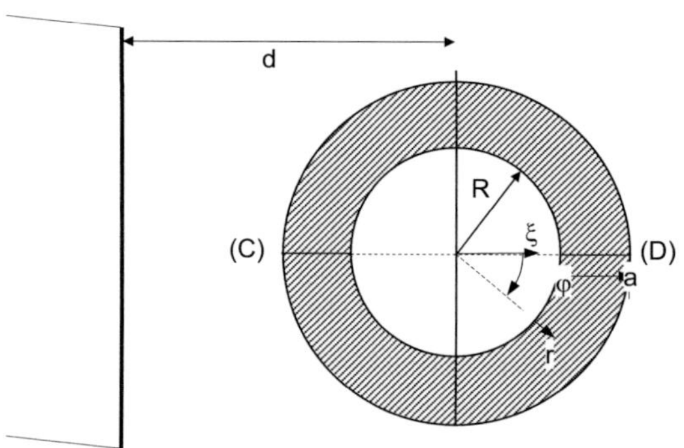

Fig. F7.3 Ring-shaped crack ahead of a straight crack.

The stress contribution to be evaluated with the weight function method is

$$\Delta\sigma = \frac{K_{appl}}{\sqrt{2\pi}}\left(\frac{1}{\sqrt{d+\xi}} - \frac{1}{\sqrt{d}} + \frac{\xi}{2d^{3/2}}\right), \quad \xi = (r+R)\cos\varphi \tag{F7.1.15}$$

The stress intensity factors for the ring crack may be computed for locations (C) and (D) as indicated in Fig. F7.3. The related angles are $\varphi=0$ for (D) and $\varphi=\pi$ for (C).
In Fig. F7.4a the total stress intensity factor K_{tot} for $\varphi=\pi$, including the total stress distribution $\sigma_0+\sigma_1+\Delta\sigma$ and the portion K_0+K_1 caused by the stress $\sigma_0+\sigma_1$, are plotted versus the relative distance from the main crack $d/(a+R)$. The stress intensity factors are scaled on K_0. The stress intensity factor terms are given as

$$K_{0,(A)} = \frac{K_{appl}}{2}\sqrt{\frac{a}{d}}\left(1 - 0.116\frac{\alpha}{1+\alpha} + 0.016\left(\frac{\alpha}{1+\alpha}\right)^2\right) \tag{F7.1.16}$$

$$K_{0,(B)} = \frac{K_{appl}}{2}\sqrt{\frac{a}{d}}\,\frac{1 - 0.36\frac{\alpha}{1+\alpha} - 0.0676\left(\frac{\alpha}{1+\alpha}\right)^2}{\sqrt{1-\frac{\alpha}{1+\alpha}}} \qquad (F7.1.17)$$

$$K_{1,(A)} = -\frac{K_{appl}}{2}\frac{R+a}{2d}\sqrt{\frac{a}{d}}\cos\varphi\left[1 - 0.401\frac{a}{R+a} - 0.065\left(\frac{a}{R+a}\right)^2 + 0.066\left(\frac{a}{R+a}\right)^3\right] \qquad (F7.1.18)$$

$$K_{1,(B)} = -\frac{K_{appl}}{2}\frac{R+a}{2d}\sqrt{\frac{a}{d}}\cos\varphi\,\frac{1}{\sqrt{1+\alpha}}\left[0.573 + \frac{0.427}{(1+\alpha)^{1/4}} - \frac{0.26}{(1+\alpha)^{3/2}}\left(\frac{\alpha}{1+\alpha}\right)^{5/2}\right] \qquad (F7.1.19)$$

$$K_{tot,(A),(B)} = K_{0,(A),(B)} + K_{1,(A),(B)} + \int_0^a h_{(A),(B)}\Delta\sigma\, dr \qquad (F7.1.20)$$

As shown in Fig. F7.4b for $\varphi=\pi$, the deviation between the approximation K_0+K_1 and the total stress intensity factor K_{tot} is less than 10% for $d/(a+R)>2$ and less than 2% for $d/(a+R)>4$. Similar plots are shown in Fig. F7.5 for $\varphi=0$.

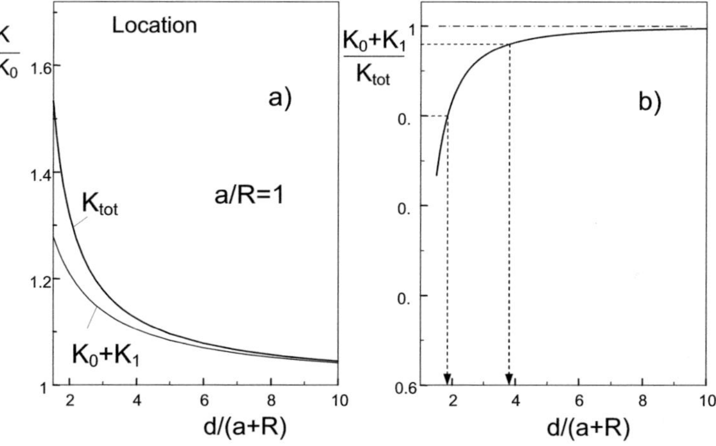

Fig. F7.4 a) Representation of the total stress intensity factor K_{tot} and the portion K_0+K_1 versus the relative distance from the main crack for $\varphi=\pi$, b) ratio of the two-terms approximation K_0+K_1 and the total stress intensity factor.

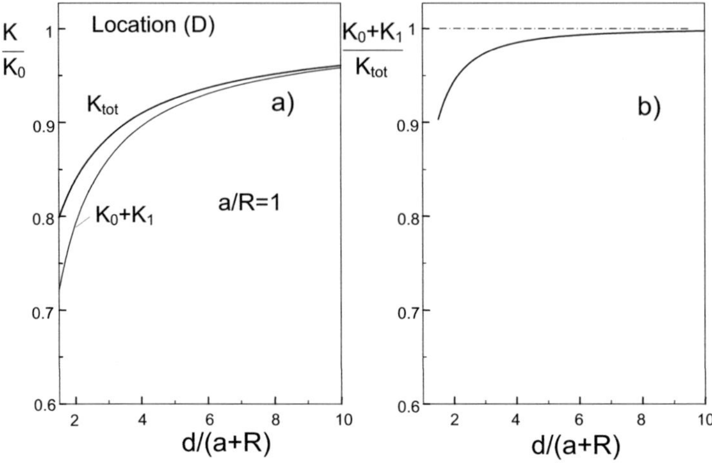

Fig. F7.5 a) Representation of the total stress intensity factor K_{tot} and the portion K_0+K_1 versus the relative distance from the main crack for $\varphi=0$, b) ratio of the two-terms approximation K_0+K_1 and the total stress intensity factor .

F7.2 Circular crack under non-symmetric stress

F7.2.1 Stress intensity factors for a circular crack ahead of the main crack

A circular crack of radius a in distance d from the tip of the main crack (Fig. F7.6a) is loaded by an applied stress σ_{appl} that varies over the diameter as

$$\sigma_{appl}(r) = \frac{K_{appl}}{\sqrt{2\pi r}} \Rightarrow \sigma_{appl}(x) = \frac{K_{appl}}{\sqrt{2\pi(x+d+a)}} \tag{F7.2.1}$$

where K_{appl} is the actual stress intensity factor for the main crack. The stress intensity factors at locations (A) and (B) can be computed from

$$K_A = \frac{1}{\sqrt{\pi a}}\left[\sqrt{2a}\int_{-a}^{a}\frac{\sigma(x)}{\sqrt{x+a}}dx - \int_{-a}^{a}\sigma(x)dx \right] \tag{F7.2.2}$$

$$K_B = \frac{1}{\sqrt{\pi a}}\left[\sqrt{2a}\int_{-a}^{a}\frac{\sigma(x)}{\sqrt{a-x}}dx - \int_{-a}^{a}\sigma(x)dx \right] \tag{F7.2.3}$$

as given by Tada ([F7.1], page 24.12). Under loading by a crack-tip stress field the stress intensity factor at point (A) is larger than at point (B).

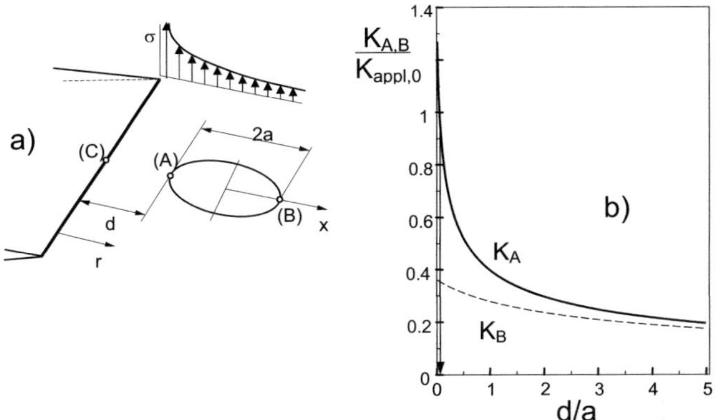

Fig. F7.6 a) Circular microcrack ahead of the main crack tip, b) stress intensity factors for locations (A) and (B), vertical arrow indicates the ratio $d/a=0.066$, for which $K_A/K_{appl}=1$ is fulfilled.

Introducing (F7.2.1) into (F7.2.2) and (F7.2.3) yields

$$K_A = \frac{K_{appl}\sqrt{2}}{\pi}\left[\sqrt{\frac{d}{a}} - \sqrt{2+\frac{d}{a}} + \frac{1}{\sqrt{2}}\ln\left(1+4\frac{a}{d}+\sqrt{8a/d}\sqrt{1+\frac{2a}{d}}\right)\right] \qquad (F7.2.4)$$

$$K_B = \frac{K_{appl}\sqrt{2}}{\pi}\left[\sqrt{\frac{d}{a}} - \sqrt{2+\frac{d}{a}} + \frac{1}{\sqrt{2}}\left(\frac{\pi}{2}-\arctan\frac{1-2a/d}{\sqrt{8a/d}}\right)\right] \qquad (F7.2.5)$$

These stress intensity factors are plotted in Fig. F7.6b versus the normalised distance d/a. For cracks in larger distance ($d{\gg}a$) it follows simply

$$K_A = K_B = \frac{\sqrt{2}}{\pi}\sqrt{\frac{a}{d}}K_{appl} \qquad (F7.2.6)$$

F7.2.2 Effect of a circular crack on the applied stress intensity factor

A microcrack in front of the main crack must also influence the stress intensity factor of the main crack. This situation was studied by Karihaloo and Huang [F7.2]. The stress intensity factor K_C at the nearest distance (point C in Fig. F7.6a) is plotted in Fig. F7.7a as a function of the normalised distance d/a of the microcrack. The data can be fitted by

$$K_C = K_{appl,0}\left(1 + 0.056\left(\frac{a}{d}\right)^{2/3}\right)$$ (F7.2.7)

where $K_{appl,0}$ is the applied stress intensity factor if no microcrack exists in front of the main crack. From the diagram Fig. F7.7a it can be concluded that the stress intensity factor of the main crack is only changed for more than 10% if $d/a<0.2$.

The normalised stress intensity factors $K_A/K_{appl,0}$ and $K_B/K_{appl,0}$ are plotted in Fig. F7.7b. The upper limit curve is computed under the assumption that the circular crack is influenced by the stress intensity factor K_C. The lower limit curve is computed assuming the externally applied stress intensity factor $K_{appl,0}$ as the loading parameter. In reality the true solution must be between the two limit cases. The stress intensity factor K_C is the maximum value along the front of the main crack. In larger distance, the K-value tends to $K_{appl,0}$. In order to minimize the uncertainty, we propose for the true applied stress intensity factor K_{appl}

$$K_{appl} \cong K_{appl,0}\left(1 + 0.03\left(\frac{a}{d}\right)^{2/3}\right)$$ (F7.2.8)

This value has to be introduced in eqs.(F7.2.4) and (F7.2.5).

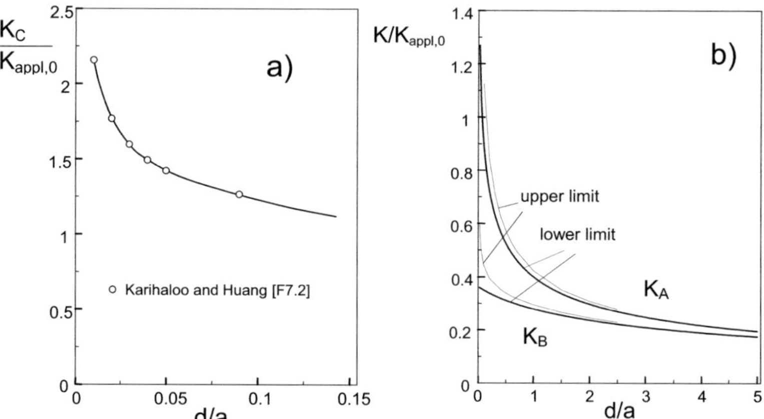

Fig. F7.7 a) Influence of a microcrack on the stress intensity factor K_C at the front of the main crack, b) upper and lower limits for the stress intensity factors at the circular crack.

F7.3 Vickers indentation crack

F7.3.1 Vickers indentation crack as a ring crack around an expanding sphere

Below the contact area of a Vickers indenter pressed into the surface of a brittle material, a residual stress zone remains after unloading. In order to determine the stress intensity factor or the crack opening behaviour of Vickers indentation cracks under the residual stresses, a description by an expanding cavity is commonly used.

For the pressure distribution it was used as a special case of the stress solution by Hill [F7.3]

$$\sigma_{res}(r) = \begin{cases} -p & r < b \\ \frac{1}{2}p(b/r)^3 & r > b \end{cases} \qquad (F7.3.1)$$

($-p$ is the maximum pressure in the centre, see Fig. F7.8). Computations of crack opening displacements were performed in [F7.4, F7.5]. In these studies, the crack was modelled by a half penny-shaped surface crack Fig. F7.9a.

As a second possibility of describing the Vickers crack problem, let us model this crack as a ring-shaped crack around the central compressive zone. The inner radius of the ring-shaped crack, R, is not necessarily identical with the size b of the compressive zone Fig. F7.9b.

If a material without an R-curve behaviour and subcritical crack growth is considered, the inner and outer ring crack radii must result from the condition

$$K_{(A)} = K_{(B)} = K_{Ic} \qquad (F7.3.2)$$

where K_{Ic} is the fracture toughness. As a consequence of (F7.3.2), the inner crack front must slightly "grow" into the compressive zone.

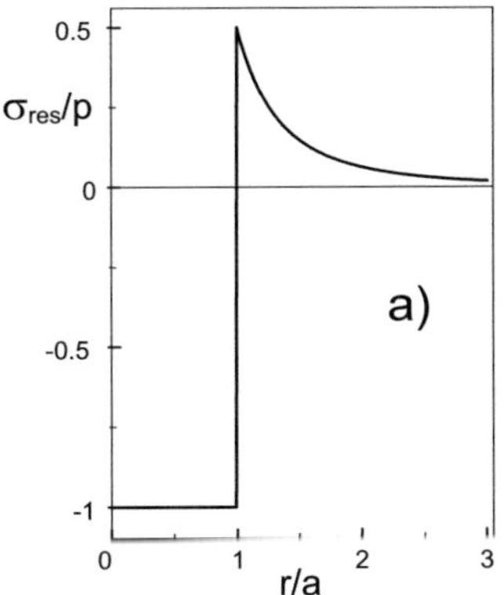

Fig. F7.8 Model for the residual stresses.

127

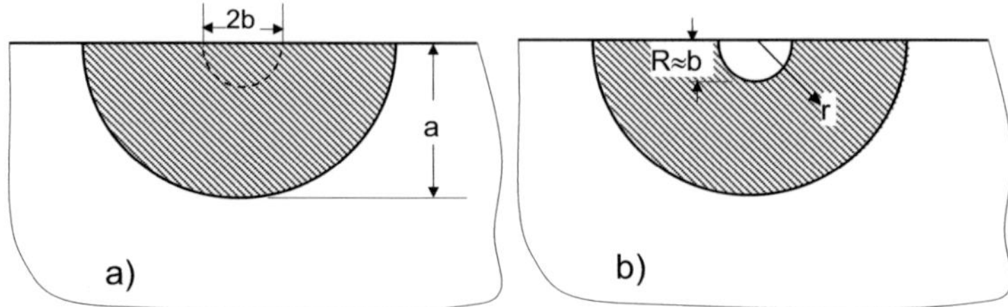

Fig. F7.9 Possible crack types to be in agreement with microscopic observation, a) half of a circular crack, b) half of a ring crack.

The stress intensity factors $K_{(A)}$ and $K_{(B)}$ were computed by inserting the stress distribution (F7.3.1) into (D2.1.1) and (D2.2.2) and using the weight functions according to (D2.1.3) and (D2.1.4). The stress intensity factor computations have to be carried out iteratively for any fixed ratio of a/R and varying R/b until the condition (F7.3.2) is satisfied. The total stress intensity factor at the outer crack contour is plotted in Fig. F7.10a versus the crack size.

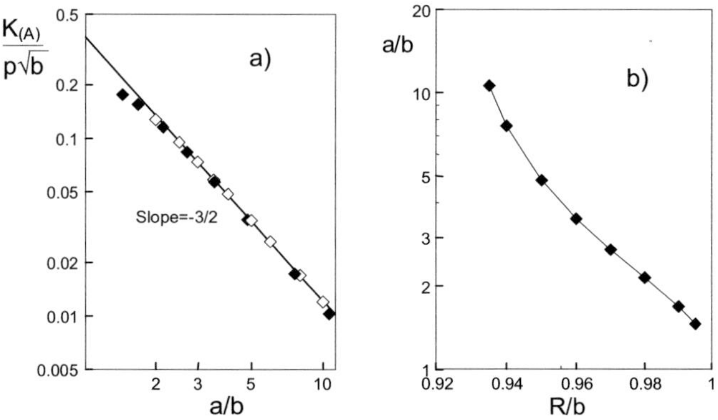

Fig. F7.10 Stress intensity factor for Vickers indentation cracks, a) comparison of results based on the assumption of a penny-shaped crack (open symbols) with results for a ring crack (full symbols), b) change of the inner radius of the ring crack.

The results for the ring-shaped crack are represented by the solid symbols. The open symbols are results obtained in [F7.4, F7.5] under the assumption of the crack being a semi-circular one loaded in the centre region by the compressive stresses of (F7.3.1). There is an excellent agreement of the stress intensity factors resulting for these strongly different crack assumptions. For $c/b{\geq}2.5$, the fitting line reads

$$K = 0.385 p\sqrt{b}\left(\frac{b}{c}\right)^{3/2} \tag{F7.3.3}$$

The larger a crack is, the more "grows" the inner crack front into the compressive zone. This is shown in Fig. F7.10b.

F7.3.2 Crack opening displacements for Vickers cracks

A Vickers indentation crack in a brittle material without R-curve behaviour is schematically shown in Fig. F7.11. The actually present applied stress intensity factor K_{appl} after unloading is related to the applied displacements δ_{appl}. An analytical solution for the COD of Vickers indentation cracks was given in [F7.4], namely

$$\delta_{appl} = \frac{4K_{appl}\sqrt{a}}{0.382\,\pi E'}\left(\frac{a}{b}\right)^2\left[\frac{b}{2a}g_2(a,b,r) + (0.635+0.319 b/a)g_1(a,b,r) - g_1(a,\lambda b,r)\right] \tag{F7.3.4}$$

with

$$g_1(a,b,r) = \sqrt{1-\left(\tfrac{r}{a}\right)^2}\left(1-\sqrt{1-\left(\tfrac{b}{a}\right)^2}\right) + $$
$$+\tfrac{r}{a}\left[\mathbf{E}\left(\left(\tfrac{b}{r}\right)^2\right) - E\left(\arcsin\tfrac{r}{a},\left(\tfrac{b}{r}\right)^2\right) - \left(1-\left(\tfrac{b}{r}\right)^2\right)\left(\mathbf{K}\left(\left(\tfrac{b}{r}\right)^2\right) - F\left(\arcsin\tfrac{r}{a},\left(\tfrac{b}{r}\right)^2\right)\right)\right] \tag{F7.3.5}$$

$$g_2(a,b,r) = \tfrac{b}{r}\left[\mathbf{E}\left(\left(\tfrac{b}{r}\right)^2\right) - E\left(\arcsin\left(\tfrac{r}{a}\right),\left(\tfrac{b}{r}\right)^2\right)\right] \tag{F7.3.6}$$

where $\mathbf{E}$ and $\mathbf{K}$ are the complete and $E(\ )$ and $F(\ )$ the incomplete elliptical integrals. The elastic modulus E' may be represented by the plane stress Yong's modulus E, since plane stress conditions prevail at the free surface.

The parameter λ can be expressed by

$$\lambda \cong 0.9828(a/b)^{0.00565} \tag{F7.3.7}$$

This solution is shown in Fig. F7.11b by the solid curves. In [F7.6] a simplified series expansion for the displacements was proposed. For $a/b > 1.4$, this approximation is represented by a series expansion. Considering the leading terms exclusively, it holds

$$\frac{\delta_{appl}}{K_{appl}} = \frac{\sqrt{b}}{E'}\left(\sqrt{\frac{8}{\pi}}\frac{x}{b} + A_1\left(\frac{x}{b}\right)^{3/2} + A_2\left(\frac{x}{b}\right)^{5/2}\right) \tag{F7.3.8}$$

with the coefficients A_1 and A_2 fitted as

$$A_1 \cong 11.7\exp[-2.063(a/b-1)^{0.28}] - \frac{0.898}{a/b-1}, \tag{F7.3.9}$$

$$A_2 \cong 44.5\exp[-3.712\,(a/b-1)^{0.28}] - \frac{1}{(a/b-1)^{3/2}} \qquad\text{(F7.3.10)}$$

A representation of the displacement approximation (F7.3.8) is given in Fig. F7.11b by the dashed curves.

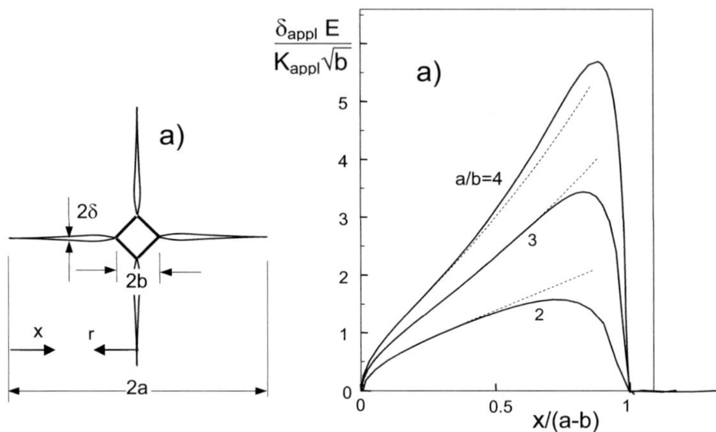

Fig. F7.11 a) Vickers indentation cracks (geometric parameters), b) crack opening displacements of Vickers indentation cracks: comparison of the analytical solution (solid curves) with the approximation eq.(E7.3.8) (dashed curves).

F7.3.2 Determination of K from the COD profile

As an example of application, the crack tip stress intensity factor K_{tip} may be determined for a soda-lime glass (E=71 GPa). In Fig. F7.12a crack opening displacement measurements are plotted as circles. The results were measured at an indentation crack introduced under 50 N load by using a SEM [F7.4]. In order to avoid subcritical crack growth during the measuring time span, the specimen was suspended for 1 h in air after indentation.

In Fig. F7.12b the measured crack opening displacements δ_{meas} are plotted versus the displacements δ_{comp} computed with eq.(F7.3.4) for a stress intensity factor of $K =$ 1MPa$\sqrt{m}$. A least-squares fit of the linear dependency yields K as the slope of the straight line, in the present example resulting in $K = 0.38$ MPa$\sqrt{m}$. Use of this value then yields the solid curve introduced in Fig. F7.12a. The dashed line shown in Fig. F7.12a corresponds to the Irwin solution for the near-tip displacement field, eq.(E7.3.5) at the same stress intensity factor. This value is roughly identical with the threshold value K_{th} in air, below which no subcritical crack growth occurs.

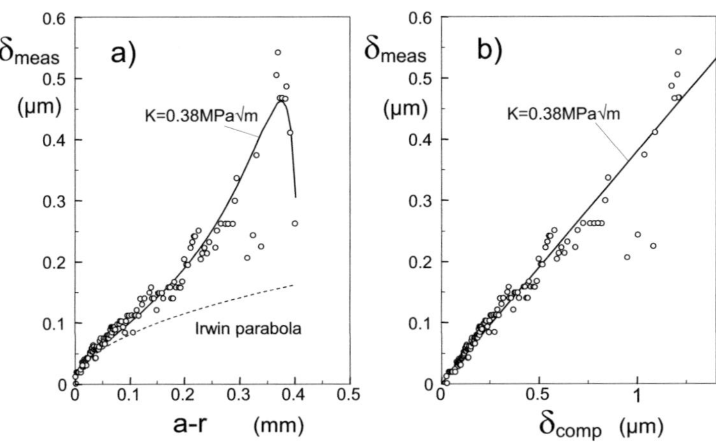

Fig. F7.12 Determination of K_{tip} for a soda-lime glass, a) measured crack opening displacement, b) measured COD plotted versus computed COD (K=1 MPa√m).

References F7

F7.1 Tada, H., Paris, P.C., Irwin, G.R., The stress analysis of cracks handbook, Del Research Corporation, 1986.

F7.2 Karihaloo, B.L., Huang, X., Asymptotics of three-dimensional macrocrack-microcrack interaction, Int. J. Solids Struct. **32**(1995), 1495-1500

F7.3 Hill, R., Mathematical Theory of Plasticity, Oxford University Press, 1950, Oxford.

F7.4 Fett, T., Kounga Njiwa, A.B., Rödel, J., Crack opening displacements of Vickers indentation cracks, Engng. Fract. Mech. **72**(2005), 647-659.

F7.5 Fett, T., Rizzi, G., Problems in fracture mechanics of indentation cracks, Report FZKA 6907, Forschungszentrum Karlsruhe 2003.

F7.6 Schneider, G.A., Fett, T., Computation of the stress intensity factor and COD for submicron sized indentation cracks, J. Ceram. Soc. of Japan, **114**(2006), 1044-1048.

PART G

CORRECTIONS IN VOLUME <u>IKM 50</u>

Page 93, Eq.(C2.1.6): It should read

$$\beta = -0.46897652 \tag{C2.1.6}$$

instead of

$$\beta = -0.6897652$$

Pages 100/101: In Table C2.5 the data for the slant crack ($c_2/a=1$) have to be divided by $\sqrt{\cos\varphi}$ (corrected data: bold numbers).

Table C2.4 Stress intensity factors ….

c_1/a	c_2/a	$\varphi\ (°)$	$K_I/\sigma_x\sqrt{a\pi}$	$K_{II}/\sigma_x\sqrt{a\pi}$	T/σ_x
		0	0	0	1
0	1	15	**0.0928**	**-0.296**	0.9545
0	1	30	**0.400**	**-0.613**	0.9026
0	1	45	**1.056**	**-1.036**	1.169
0.9	0.1	15	0.0203	-0.0742	0.8732
0.9	0.1	30	0.0810	-0.1380	0.5284
0.9	0.1	45	0.1838	-0.1831	0.0636

Table C2.5 Stress intensity factors and T-stress under remote y-stresses.

c_1/a	c_2/a	$\varphi\ (°)$	$K_I/\sigma_y\sqrt{a\pi}$	$K_{II}/\sigma_y\sqrt{a\pi}$	T/σ_y
		0	1.1215	0	-0.526
0	1	15	**1.088**	**0.177**	-0.526
0	1	30	**0.989**	**0.329**	-0.411
0	1	45	**0.838**	**0.434**	-0.1013
0.9	0.1	15	1.087	0.1696	-0.3805
0.9	0.1	30	0.989	0.3172	0.0002
0.9	0.1	45	0.838	0.4255	0.4985

Table C2.6 Stress intensity factors and T-stress under constant internal pressure p.

c_1/a	c_2/a	φ (°)	$K_I / p\sqrt{a\pi}$	$K_{II} / p\sqrt{a\pi}$	T / p
		0	1.1215	0	0.474
0	1	15	**1.179**	**-0.119**	0.544
0	1	30	**1.387**	**-0.285**	0.804
0	1	45	**1.893**	**-0.602**	1.484
0.9	0.1	15	1.108	0.0951	0.4926
0.9	0.1	30	1.070	0.1792	0.5309
0.9	0.1	45	1.022	0.2424	0.5616

Page 106: In eqs.(C3.1.10) and (C3.1.11) the angles β agree with angles φ.

Page 110: In Fig. C3.3 the designations of the g_{ij} and C_{ij} have to be changed: corrections bolt.

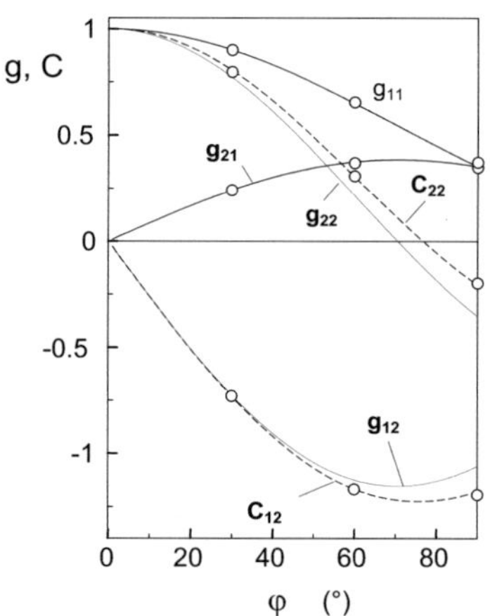

Fig. C3.3 Stress intensity factor ….

Page 275-277: In eqs.(C15.2.1b)-(C15.2.1e) the variables b/R have to be understood as $|b/R|$. The same holds for eqs.(C15.2.3b)-(C15.2.3e).

Equation (C15.2.2) doesn't show the correct behaviour for very small misalignments which should satisfy $dF/d(b/R)=0$ at $b/R\to 0$. Therefore, a quadratic fit with respect to (b/R) was performed. For the result see eq.(E5.5.2) in this volume

$$F_I = F_{I,0}\left[1+\left(0.122-0.003\frac{a}{R}\right)\left(\frac{b}{R}\right)^2\right]$$

(E5.5.2)